From earliest childhood Susanne Hart wanted to be a veterinarian. She studied at the Royal Veterinary College, London where she met her husband, Toni Harthoorn. She has written two previous books, *The Tame and the Wild* and *Life With Daktari*.

After she finished writing *Vet in the Wild*, the Harthoorns and their children moved to Southern Africa, Toni, a pioneer in the large wild animal immobilization, having joined the Department of Nature Conservation there.

VET IN THE WILD

Susanne Hart

Fontana/Collins

First published as *Listen to the Wild*
by William Collins Sons & Co Ltd, 1972

First issued in Fontana 1977

Made and Printed in Great Britain by
William Collins Sons & Co Ltd, Glasgow

CONTENTS

ACKNOWLEDGEMENTS

I wish to express my most grateful thanks to my husband, Toni Harthoorn, for his help and patience as well as for his photographs; to Jock Gibb for his continuous encouragement; to George Adamson for sharing with us his immense knowledge and love of the wilderness; to Nicky, Liz, Jean, Tom, Gail and all those who so kindly assisted with both manuscript and photographs; and to the many who have directly or indirectly made this book possible through their untiring efforts in the cause of wildlife.

INTRODUCTION

So much has already been said and forgotten about wild animals; they have been photographed, filmed, studied, eulogized, hunted and enslaved. Each year tourists flock to view them in their habitat, many still wrapped in their own world of yesterday, unable to lose their identity in the magnificence and stillness that surrounds them. Africa welcomes, of necessity, not only the admiring tourists but also the hunters who are bent on destroying the very creatures that enthral them, only content to return home laden with trophies.

Yet among those who come to seek out the African wilderness there are many who regard the Parks and reserves as sacred, inviolate refuges of Nature, whose wild magic recharges their stress-battered bodies and souls. Some come for a week and stay for many months, unable to face again the tedium of their previous existence.

During the twenty years my husband Toni and I have been privileged to work in Africa our lives have been interwoven with the wild. As veterinarians, with Toni's exceptional scientific knowledge, our work ranges from investigation and research to veterinary treatment of animals such as the elephant and giraffe, down to the lesser wild cats, small antelope and mongoose. Life is too busy to record all our daily and hourly observations in detail; yet some of the remarkable traits and behaviour patterns of the wild animals placed in our care could not be permitted to drift into oblivion.

'What a life you lead,' said a young bookshop assistant whom I met during a promotion tour in England for my previous book, *Life with Daktari*. 'I'd do anything to come and help you with the animals, if only I could find a way to get there.'

'Writing another book?' one enthusiastic writer asked

me, 'about conservation? That word,' he said, warming to his theme, 'is much too hard for the man in the street to understand. Saving whole species, whole masses of wild animals, is just too big, too wide a concept to carry any appeal. But when you write about the individual animal, like that cheetah cub with the injured legs, or the orphan rhino, now that's a language readers can understand. And don't forget, your world is strange and new for most people, so that reading about your patients and adventures, one at a time, builds up the picture slowly, so that we can live it all with you. That way you get right into our hearts until we feel that the wild animals are our responsibility too.'

Vet in the Wild was born from that thought. Our patients, as different from each other as one human being from the next, are also as varied in their response to us, each other, their environment and their predicaments as the multitudinous shades of the African bush. At times we must go to them to make treatment most effective; yet often, too, they are brought to us so that we can give them daily and perhaps hourly care. Our contact with them may be very brief, as in the case of the snared giraffe; or very much longer, lasting weeks and sometimes many months.

However long or short a time we have spent with them, each case counts as a new experience, an adventure in itself, often leading to fresh knowledge which might in some way be put to use for others. Inevitably, because our home cannot accommodate a permanent menage of wild animals, there comes the time for parting, leaving us with a deep sense of loss and the memory of their courage and immense endurance; of their marvellous sense of frolic and mischief contrasting their intensely heartrending though often silent, manifestation of grief. Yet whatever their mood or emotion, their illness or injury, there was one need, one compulsion, contained by them all: the ardent, passionate desire to live on, at whatever cost or sacrifice, a desire which burnt within them as surely and as fiercely as it does in man.

Chapter One

THE COURAGEOUS GENET

Iain Douglas-Hamilton and Orio lived with their baby in Manyara National Park in Tanzania where Iain had almost completed four years of elephant research. Alicat the genet and his playmate Wiji the mongoose lived half wild in the beautiful camp set between Lake Manyara, home of countless water birds, and the steeply forested escarpment. Iain had chosen to study the most magnificent and enigmatic of all African mammals, but four years, the time allotted for his work, was hardly enough – he would need several lifetimes to solve their mystery.

The Park, set in 35 square miles of land and 8 square miles of lake, proved a perfect study area, for the population of 300 elephants was fairly constant. Iain had come to know the elephants individually and almost intimately; had seen them give birth, mate and fight; had witnessed the formation of new groups, had studied the breeding habits and the progress of the young and had followed their movements by means of tracking devices. The thesis which he hoped to submit for his doctorate should read like a saga.

The little genet was given just as much love and attention as were the elephants; the mongoose received as much love as did the genet and the same went for Iain and Orio's baby. Their Seychellois nursemaid complained loudly and bitterly that the parents loved the genet more than their own child, otherwise why did they shed so many tears over it when they had never once wept over the baby? Orio tried to explain to their old servant that the baby was well and so there was nothing to weep about – but that wasn't good enough. To the nursemaid, lack of tears was synonymous with neglect!

I think that we all, at some time or other, wept tears over Alicat, probably the first and perhaps the last genet that has ever been airlifted to seek veterinary attention. Poor Ali, he had been in a very bad way indeed. A savage attack by an adult male genet whose terrain he had dared had resulted in a shattered lower jaw, torn gums and muscles, as well as injury to one back leg which was paralysed together with the lower spine. In spite of the terrible damage he had suffered he looked at us with courageous defiance from his stronghold on Iain's shoulder as if he had already made up his mind that he was going to live and was prepared to fight for his life.

'Any hope at all?' Iain said despondently. 'Will he have to be destroyed?'

'Let's take a closer look at him,' I said and went to make up a cup of bland surgical solution. When I had cleaned the wounds up as much as I could I filled a syringe with diluted milk and glucose, and put a very thin polythene tube on to the nozzle. Then I injected some of the liquid into the tube and approached the apprehensive genet, still draped round Iain's shoulder and neck, and talked to him quietly for a few moments while I stroked him.

He reacted by half closing his pain-filled eyes, putting his ears forward and winding his tail more closely round Iain's neck as if to brace himself for whatever I intended, which at that moment was to test his feeding ability. I introduced the thin tube by guiding it on to the upper palate, then pressed home the plunger so that a small amount of fluid dripped into his mouth. At once Ali responded; he began to suck eagerly, his very long, spoon-shaped tongue coming into full play. Obviously the genet was both parched and starved, judging from his brittle coat and tightly tucked-in flanks. He had been injured over a week ago, but had disappeared almost immediately afterwards, perhaps to seek a place to die. Three days later, exhausted and scruffy, he had somehow crawled back into camp.

The mongoose, his friend, had shunned and repelled him as animals will even when their own young are sick or injured. Perhaps this rejection had upset Alicat more than one could guess and had induced him to relinquish his home just when he needed it most. No one will ever know where he holed up for those three days but by the time he returned his mouth had become necrotic, his eyes glazed with weariness, his back legs non-functional. Orio had unsuccessfully tried to feed him but must have touched the injured tissues, for he refused even egg, his most favourite food. Rather than let him die Iain had put Alicat into his two-seater plane and flown him to Nairobi, a flight of one and a half hours, though by road it would have taken him most of a day.

Ali's energetic response to our test gave us hope. The question now was, how to repair the minute bones and then restrain him while healing took place; that is, if it could when part of the tissues were already dead.

'Leave him with us,' Toni said, 'and we'll see what we can do. We will have to anchor the whole jaw structure with dental wire, but until we have anaesthetized him, we cannot really see the full extent of the damage. And then there is the paralysis and the question as to whether function can ever return to the vital organs in the damaged area. We'll contact you when we know a little more.'

When Iain had left, after a sad farewell from his beloved Alicat (derived from alley cat), we injected the genet with a small trial dose of tranquillizer and a wide-spectrum antibiotic. It would be best for him to live in the house, rather than put him into the cottage at the end of the lawn known as the 'cheetah house' which was both feline sanatorium and human guest room. On occasion it had been used for both purposes at the same time, depending on the tolerance of our visitors which was quite unpredictable, usually even to themselves. They might adore wild animals and might have travelled to East Africa for the sole purpose of seeing them, but even so a convalescing cheetah on or in one's bed might pose insurmountable problems.

Alicat was accommodated in our best spare bedroom which, by good fortune, was vacant at the time. As soon as the sedative took effect, we put him into his own box complete with a towel impregnated with Iain and Orio, to make him feel at home, and left him to sleep in peace.

Seeing that it was coffee time Toni and I sat down and discussed every aspect of our case, weighing up the chances, trying to decide upon a plan. We had taken on a mutilated, toxic, sub-adult genet which was unable to feed itself. This meant hourly and, if it survived, daily care for weeks at the very least.

Even though it *had* made an effort to feed, on the face of it there was little hope. Toni and I had often been faced with impossible, apparently hopeless cases and we had learnt something very vital from them: that healing and recovery are not *only* dependent upon the veterinarian's knowledge or even his skill, but also a little-understood healing force which can produce the most astonishing results.

It was the same healing energy force which we breathe in when we breathe fully and consciously and which can be drawn at will to any part of the body to alleviate pain and restore health. Few acknowledge that this natural force exists and can, indeed, be tapped and used by anyone, however sceptical or uninitiated.

Alicat was not only courageous, but had the will to live, essential in a case when so much depended on the patient's co-operation. He allowed us to befriend him almost at once and trusted us implicitly.

'It's going to be one hell of a battle,' Toni said thoughtfully as he sipped his coffee. 'If we hadn't seen miracles happen so often, I would say that there is no way for this genet ever to have even a semi-normally functioning lower jaw again. As it is, miracles or no miracles, we shall have to be on day and night duty for some time. That is, if he lives through the operation.'

'No one but you two would have taken this on,' Iain had said as he left our house to return to his elephants; but even as he said it, we could see that he did not really expect ever to see Alicat again.

That day we cleaned the broken jaw while Ali was under the influence of a tranquillizer. For the first time we could see the exact extent of the damage, which was considerable. The two rami (wings) of the lower jaw were anteriorly exposed, the fractured ends hanging downwards. As we had feared, decay had set in with accompanying discoloration caused by tissue debris, pus and dislodged bone. Apart from the mechanical difficulties which such a compound jaw fracture presented, the infection would have to be controlled and overcome if Ali was to survive.

What amazed us was that the flame of life still burnt so brightly in the slender, lithe little body, in spite of extreme stress and starvation. Unlike an adult animal with reservoirs of fat and well-built muscle from which to draw in times of illness, Alicat, immature at seven months of age, had no such reserves at his disposal.

From what secret source then, did he derive the energy which eventually brought him back to camp and safety? How did he find the strength to leap on to Iain's shoulder and purr and stretch and suck the liquid which we gave him? Did he continue to live when, medically speaking, he should have been dead because of an inborn instinct which demands life at all cost, or because the number of his days on earth were not yet complete?

Ali, breathing deeply and regularly, lay on a white cloth under the glow of the standing lamp. We had decided to operate the night after he arrived so that the infection-combating antibiotics had time to take effect and we could be certain that the tranquillizer we were using had no ill effect and would combine successfully with the general anaesthetic we reluctantly had to administer. We had no precedent to go on, no drug dosage we could follow, for the genet, feline in many ways, is not a true cat, having been classified – perhaps erroneously – as a member of the *Viverrida* family, which includes the mongoose and the civet, both possessing non-retractile claws.

'All right, let's drape him.' Toni, satisfied that the respiratory rhythm was satisfactory, picked up the magnifying lens to examine the jaw more carefully while I wetted the coat with surgical solution to keep loose hair in check. How incredibly versatile my anaesthetist/fellow surgeon was! Having pioneered tranquillization and anaesthetic techniques in the large-hoofed wild animals of Africa so that scientific conservation methods as well as relocation of endangered species had become possible, he had, more recently, established suitable dosages and drugs for use in wild feline animals, such as the cheetah, leopard and lion. This miniature semi-feline creature, weighing no more than two pounds, or possibly three when restored to maximum health, posed a very different problem and we decided to go very carefully with medication, from antibiotics to knock-out drugs.

'Take longer over it, wait for hours if necessary, repeat and repeat again in gradually increasing doses. If you haven't got the time to take it slowly, then either *make* the time or don't tackle the case at all,' Toni used to say.

We had obtained some of the very finest dental wire, hoping that enough of gum and tooth was left to use as anchorage. After cleaning up the fracture site and removing dead fragments of skin, bone and muscle and by wedging the molar teeth apart for easier approach, we were able to repair as much as possible of the damaged, but still viable, tissues and bring together the lower sections of the rami. Stooping over the tiny head we worked together as fast as we could for, according to our calculations, the anaesthetic time would be no more than half an hour, after which, we hoped, our patient should begin to recover.

All went well, except that when we had finished, surgically speaking, the injured lower jaw, now trimmed down to healthy bone, seemed terribly short. Would Alicat ever be able to feed himself again, or live a relatively normal life? 'Think positively,' Toni said, stifling my doubts, 'and all will be well.'

Almost as he spoke the words, our first worry, which concerned Ali's tolerance to anaesthetics, was allayed. He stirred under the drape, made a sound which was half mew, half growl, and moved his head. Immediately, we took him away from the enervating heat of the standing lamp, put him into his warm box with a plastic hot water bottle, made sure that his tongue was in the right place and left him in the darkness to be checked at hourly intervals throughout the night. He had tolerated all medication amazingly well, taking a dosage rate far in excess of his two-pound weight ratio. Unfortunately he recovered too quickly, for before we were fully aware of it, or had a chance to increase the tranquillizer, he had removed some of the wire sutures with his sharp retractile claws. We were utterly dismayed by this vigorous self-mutilation, realizing that only two courses of action were open to us: either to give up altogether, or to continue, which would involve another operation using a different technique and stronger dental wire. We chose the latter, and again, two days later, Alicat responded marvellously. Although resurgery meant a definite set-back to tissue healing, especially as in his efforts to undo ours he had done further damage to himself, it did, at least, give Toni another chance to re-test the efficacy of his first genet anaesthetic procedure.

After the second operation we kept Alicat more heavily sedated, so that for three days he was a very subdued genet cat indeed, emerging only when I came in with the feeding tube and retreating into his up-ended box immediately after his three-hourly meal. I had, naturally, to relieve his internal pressures artificially since his post-spinal paralysis, almost certainly due to a blow from the wild genet, was still very much in evidence.

Although his lower jaw was terribly fore-shortened, at least it all looked clean and firmly anchored into place, with pieces of shining metal and suture material interspersed between what was left of the normal tissues. In profile Alicat looked very strange indeed, rather like someone with a protuberant chin; so from this angle it was impossible to see that almost half of his lower jaw structure was non-existent. From the top more usual view he looked perfectly normal, except that his face appeared 'drawn' in the same way that human features alter under extreme stress. His coat lacked sheen and suppleness and his movements the elasticity he was so soon to regain. Apart from this, his demeanour, especially after the first post-operative week, was that of a playful purring adolescent who hadn't a care in the world.

On the fifth day after the second operation, two remarkable things occurred: first, he regained the use of his right back leg and his posterior organs. The former recovered gradually, with a slight limp persisting for a few more days. This slowed him down on his up-hill vaulting escapades, doubtless a good thing, considering how fragile he must have been at this stage. The posterior organs now functioned normally.

Secondly, and most unexpectedly, he began to lap by himself when, on impulse, I offered him the saucer of fortified milk, egg and liver juice from which I was about to tube feed him. His mouth and jaw must have been very tender indeed, his stitches causing permanent discomfort, especially when he yawned, which he did often and with great abandon. Yet we could only guess that this was so, for as soon as he felt his new independence and could eat whenever he felt like it from an ever-present bowl, he became very active, especially after his long sleep, which lasted most of the day into early evening.

By the time we removed the last of the stitches two weeks after surgery, he was chewing small pieces of cooked liver with relish. His tongue seemed to take the place of the anterior lower jaw whose two sections were miraculously knitting but still palpably mobile. He would curl his tongue round the semi-liquid food, throw it into his mouth like a performing seal and chew with his still-efficient armoury of molars. On occasion he would bring back the homogeneous mass with one shake of his enchanting head and start again, either because it was fun or else because his food wasn't macerated sufficiently. After a daytime meal he would retreat to rest and to digest. If it happened to be evening or night, he would leap up on his favourite curtain rail and lick himself with intense care rather in the manner of a domestic cat, except that his serpentine tail came into full play on such occasions, helping him to balance on the precipitous curtain-rail ledge. Twenty-four days after we had taken on

Alicat as a patient it became obvious that his shattered jaw was healing in a manner which could no longer be solely attributed to medical or surgical care.

We had been on safari for a night and two days, leaving our household in the tender care of our able African maid Jenny and our friend Jean Hayes who had promised to look in and give comfort to our dogs and patients. As soon as we arrived home we entered Ali's den where he was just in the process of awakening from his long daytime sleep. I scratched his elegant long neck for a few moments, then let him leap on to my shoulders. As he did so, I quickly and very firmly took him by the scruff for the spraying operation for which we used an antibiotic gentian deep-penetrating spray to keep dreaded infection at bay. As soon as Ali felt my fingers on his neck he went limp – perhaps realizing that non-resistance was the better part of valour and it would all be over much more quickly if he kept still.

As soon as the blue cloud subsided – its bitter vapour in our mouths – and Toni gave the signal, I hoisted the brave genet on to its high place and presented it with a titbit. Each time he responded in exactly the same way: he looked down at us disdainfully ignoring the peace offering and our shoulders on to which he would have leapt immediately at any other time. He then wiped his jaw and face with one long, lithe, sensuous movement all along the white curtain pelmet, until he felt content that he had, in some small measure, given us back a little of our own medicine. This ceremony complete, he jumped on to the feeding platform with one perfectly measured movement and proceeded to tuck into his bowl of food. After that, if we did not escape quickly enough, he would catapult on to our necks like lightning and mark us, too, with the purple dye! Toni put on the strong desk-light and I held Ali up for inspection: the deep pink of newly granulating tissue had moved forward and across.

'What's this?'

Toni, with a look of incredulity on his face, was pointing to the new advancing hair line which certainly had not been there before we left. Hair was overgrowing the new scar tissue which normally is devoid of hair follicles and therefore remains bare.

'Incredible,' he said. 'I would never have believed it. Do you think the whole jaw will regrow in time?'

I was about to make a rejoinder when Alicat, suddenly released, became tangled up in my hair which he treated like a ball of wool. The fact that he only possessed half a lower jaw seemed to make very little difference to his effervescent spirit. He had already overcome the seemingly insurmountable problem of how to feed himself and would no doubt solve any further difficulties in the same indefatigable way.

'The sky is the limit for Alicat,' I answered as soon as I had extricated him from my hair. 'It wouldn't surprise me in the least if he grew a new set of teeth too by sheer willpower.' Our patient's energy and *joie de vivre* increased greatly as his jaw continued to heal. My typewriter became his favourite rubbing post, the moving carriage a special source of amusement and acrobatics. He swung up on to the curtains with an ease which would have made a trapeze artist seem like a snail by comparison. He levitated rather than jumped, seeming to use his multi-purpose tail as a springboard; only a slow-motion film could have caught the secret of his inner-spring mechanism; our eyes were much too inept and always several movements behind.

Ali played with the simple toys we made him: a rolled-up piece of newspaper attached to the curtain rail, or an ostrich feather, which he attacked ferociously for hours at a time, and which no doubt conjured up the image of a prey animal.

While I tried hard to concentrate on serious work at my desk, Ali swept from room to room like a tornado. He adored the children's study-bedroom wing most of all since it contained their sea-shell collection and countless other treasures like bird feathers, weavers' nests and even a sloughed snake skin. He attempted to reach this special heaven by taking a running jump, in best Olympic style, and landing neatly on the door handle to force it down. Luckily for us his light body did not make much impression, though had he stayed with us for long enough and maintained his weight increase he would certainly have moved that obstinate piece of metal by sheer persistence. As it was, he just kept trying: we knew because we heard the metallic click emanating from his room from time to time.

During the day Alicat disappeared into Toni's old long-neglected boxing-gloves which I had mounted, two tiers of boxes high, on the spare guest-room wardrobe. When, in the morning, I came in to see if all was well with him and to give him food and water, I would be lucky if he favoured me with so much as a half-pricked ear or a twitching whisker before he once more withdrew into the fastness of his padded castle.

One week after the new hair-growth had first appeared Orio came to pay us a call. She and the baby were now staying on her parents' farm on the shores of Lake Naivasha where Orio was making arrangements for their imminent move to England. We were not aware that anyone had arrived until we heard our smaller dog Jessie vocalize enthusiastically, after which the tinkle of our Greek bells suspended at the hall entrance announced the arrival of a guest.

I called out but there was no reply. We had been sitting in the fading sunset drinking late tea and were loath to switch on a light until we had to. Just then we heard a soft footfall and as we peered into the half-darkness we saw a cloaked figure advancing slowly towards us. And still we did not guess for, to our call of 'come in and have some tea, whoever you are', there had been no reply at all. Toni got up and switched on the light and only then did we recognize Orio, a glowing embroidered shawl flowing from her shoulders, her arms stretched out towards Alicat, who was practising his acrobatics on the typewriter keys. Deeply moved at the thought of seeing her genet again, she had at first been quite unable to utter a word; she kissed us effusively in turn, and then stood very still next to Ali, tears pouring down her cheeks, until he recognized her and leapt on to her shoulders. Purring and stretching in ecstasy at having found his long-lost friend, he snuggled into her long hair and stayed there.

'But surely, Toni and Sue,' she said with her husky, captivating Sophia Loren voice, 'surely that jaw has grown. Is such a thing really possible?'

'You can't stop tissues from healing,' Toni smiled; 'all we can do is to help them along and give them a chance.'

Orio nodded. She seemed to understand that there was more in heaven and earth than could be explained with science and left it at that. For a while longer she played with Ali who had no intention of leaving her now that he had found her again. Finally, after much coaxing and tempting with some of his favourite titbits, he managed to extricate himself from her long, chestnut tresses to take the food. With a long, last, loving look at her genet she slipped away before he knew she had gone.

In spite of all our precautions, Alicat did escape once ten days before he was due to leave and just after Orio had thanked us for keeping him safe. It happened one night, when he must have slipped out of the window that we found open the next morning. To me this was a disaster, to Toni less so, for he was convinced that Ali wasn't far away and would return, just as our previous genet, Poppet, had done.

After searching for him everywhere in the garden and the house – he could have been shut into a cupboard or drawer very easily – we put out the word that our genet was at large and offered a reward for any information. I left tempting bowls of food in every conceivable place hoping that Alicat would return home when he was hungry and thirsty. I opened windows in every room in the house so that he could get in when he was ready to do so.

On the morning after the second night of his disappearance the mousebirds and black-headed weavers which inhabit our garden made a tremendous fuss in the thickly covered tree-crown where they build their nests. They hopped up and down as if they were mobbing an intruder, the way I have seen them behave when a snake is near. On the other side of this thicket was the neighbour's garden and in it two enormous, ferocious-looking cats and several dogs. Unlike our own two dogs who tolerated any kind of strange animal in their home providing it was part of our 'family', other dogs might not be so kind and if they killed our genet one could not really blame them.

'Let him be and don't try and chase him out,' Toni suggested. 'He is bound to be up there and will appear when he is hungry.'

We spent a tense three days, a long time for a convalescent immature non-city-orientated genet to be at large.

On the fourth day, exactly as Toni had predicted, Alicat came out of hiding, climbed up a small tree next to the hedge which was in the neighbour's garden, and was spotted by the gardener who informed Jenny who brought him home. We were out of Nairobi that day; to come home and find Ali on the pelmet again was almost too good to be true.

The fact that he survived three days and four nights – and survived well for, apart from being leaner than usual, he was unchanged – proved that his recovery was now almost complete. But more than that: he had proved that he could now fend for himself; his escape, hair-raising though it was, acted as a kind of trial run for the future when he would have to be exposed to the elements even if he did have the security of a shelter and a good meal to come home to.

Chapter Two

SEVEN DAYS WITH A MONGOOSE

Everyone must have a place to call his own, genet, mouse-bird, elephant or human. Home is not necessarily a house, a nest or a burrow: many wild animals, and especially the large-hoofed gregarious ones such as the elephant and the giraffe who eat almost continuously to keep alive, move vast distances to feed, finding security and protection within a group, or herd, or even in the company of a single mate. The larger cats, like the cheetah and lion, have vast home ranges, which they guard fiercely. Even so, the eternal search for prey often drives them beyond the edges of their own territory, as in the case of the lions of the Serengeti plains, some of which follow the prey animals when they begin their vast migration.

Toni and I love to travel to distant remote places but when we have had our fill there is nothing more wonderful than to come back home. Almost always, even after only a short absence, someone has been there before us. They may have come and gone, unable to wait any longer, leaving a message scrawled on a pad: 'lion much better' or 'come as soon as possible, ostrich very sick', or 'why weren't you here, we had a date!' or, 'couldn't wait any longer so left the leopard cub with you, it was born this morning and is injured . . .', or, 'my baby is constipated again, will call tomorrow' (no mention what species of baby). We have come to expect unusual messages, people and events. There are months when all is relatively quiet, and we can catch up with neglected work. It seems to hit us in waves which are neither related to season nor to anything else that we can discover. Everything usually happens at once, and when it does, it means all hands on deck: that is, ours, the children's when they are home on holiday and any other person's who is kind – and foolish – enough to offer us help.

Wiji, Alicat's playmate and our first-ever mongoose patient, arrived late one night, tucked into Iain Douglas-Hamilton's shirt.

'Another Hamilton patient,' he said, smiling apologetically; 'can you stand it? She really is terribly ill.'

The mongoose, like Alicat, had not eaten for days, but for quite a different reason. Wiji Titino, so named by her Italian mistress, had taken something very corrosive and was unable to swallow. What had actually made her so ill we could not establish. It might have been battery acid, or a poisonous plant, or a skin dressing which had been applied to her when she contracted mange. Whatever it was it had burnt its devastating way through her digestive system with a vengeance, wreaking terrible damage on the way.

'Anything you can do?'

Wiji, clinging to Iain's chest hairs with her remarkably expressive hands, did not want to be examined in the least. She struggled and squeaked and made a terrible fuss, threatening to bite us all if we did not leave her alone.

'We can only help if we can control her. Looks as if she needs sedation.' Toni worked out the dosage rate, using the same drug we had found so effective for Alicat; then he divided the amount into half, taking into account the mongoose's weak state and the fact that the cause of her illness was not really known. While Iain held her tightly for a moment the hypodermic needle directed the sedative dose into her quivering, outraged body, amazingly resistant in spite of her condition. She clung to Iain until, at last, half an hour later the sedative took effect and she relaxed, her fear of being parted from him wholly dispelled.

'Thank heavens for tranquillizers, with a wide margin of safety,' I thought, as I eased the limp, stiff-haired little body into a box of straw. By the time the drug effect wore off, long after Iain had left, she should be calmer. Meanwhile I took advantage of her sleepiness and medicated her for acute gastro-intestinal inflammation, which included sub-cutaneous injection with saline solution to prevent dehydration. The skin condition, almost certainly parasitic, would have to wait. The main problem was to restore the epithelium of her mouth, her throat and her digestive tract.

'How about Rabies, there was a strange dog in the camp which we destroyed,' Iain confided just before he flew off, which left us bewildered and speechless, at least until we were able to reconstruct the sequence of events.

It seemed that Orio and Iain, contrary to previous custom, had left Wiji with their head-man as they boarded their plane which was kept next to camp. Desperate at being rejected, the devoted mongoose caught them up and leapt on to the wing of the plane, her usual way of boarding. But neither Iain at the controls nor Orio holding her baby could accommodate a very fat (as she was then), possessive, sharp-clawed mongoose as well as everything else. She was peeled off the wing and taken back to camp amidst angry squeals and violent struggles, but Iain was firm and Wiji, whom they both adored, had to stay behind.

When they returned two weeks later, Wiji had changed; she was thin and unhappy, her body and tail-coat no longer burnished red-brown and sleek. She made only a token gesture at eating and drinking, going steadily down-hill even though every effort was being made to treat her. At this time the strange dog had appeared as if from nowhere and had played with Wiji. He was a sick-looking animal covered in mange, so Iain destroyed him. Instead of a happy homecoming, their return was shrouded in gloom, especially as they held themselves responsible for Wiji's illness.

The stray dog's appearance and Wiji's inability to drink made everyone think of Rabies. We tried hard to retrace the steps of her illness one by one, but too much had happened in the meantime. First had come the mange and the treatment; then she had begun to scour and had been treated for that in turn and from then on she had declined rapidly, losing her rotund shape until she was skin and bone.

The mongoose is known to be a carrier of Rabies, the dreaded disease whose first symptoms are a change of temperament and the inability to take water, though thirst is extreme. Almost everyone in camp had been bitten – or rather nipped – by the mongoose at some time or other in the not too distant, and distant past. Wiji had always been playful, and her bites, which usually drew blood, were not usually intended as a form of aggression. Her inability to swallow could easily be explained if her mouth and throat had been damaged by a strong poison; yet complete mystery surrounded its origin.

We discussed it, weighed it up carefully and kept any real worry to ourselves. Toni and I were sure that Wiji did *not* have Rabies: unless it was a suppressed case, she would be dead in a week after the first symptoms appeared and she had already been off-colour for longer than that. Her change of temperament was surely due to pain, frustrated hunger and thirst and her inability to do anything about them.

To our surprise a blood test revealed that Wiji, apart from all her other troubles, also suffered from tick-fever. She had developed a touch of pneumonia and for two days she hovered between life and death.

After three days of artificial feeding she gave her first sign of life by chirruping a welcome as I came in to tempt her with breakfast. Liquefied pawpaw did get her on to her feet for a moment, though she did no more than wet her snout with it. Like Alicat, she loved banana and egg but shunned milk, probably because her digestive system could not yet cope with it. When I left the bathroom that morning she managed to climb out of her box for the first time. We transferred her on to the lawn during our breakfast hour, when we could watch her closely.

From then on she improved steadily, making frequent forages into the garden in search of insects which as yet she could not swallow. I now began to forcefeed her hoping to stimulate her appetite, maintaining regular vitamin B injections as well as other medicines. Her small, prickly-coated body was only just holding its own, but her spirits were improving by the hour. Small quirks of behaviour, such as testing the bottom of her food bowl with her long-toed paw, meant interest and returning strength. While she was critically ill she did not chirrup once, but gave a series of high-pitched mongoose-squeaks whenever we were with her. When she began to recover, her chirrups – very cheetah-like – came in a stream at equal pace with the squeaks – and there was never silence, except when she slept.

Her sleeping posture also gave us a good hint of the state of her abdominal and lung affliction: when she was at her worst she rested, her front legs stiff, her head drooping forward into the soft bed of grass (to relieve pressure from her lungs), so that she never really relaxed in sleep. As soon as she improved she slept once more with her head tucked under and backwards, so that she looked more like a rufous miniature hedgehog rolled into a ball, rather than a mongoose, a mammal belonging to the genet and civet family, in spite of its unfeline appearance.

Eventually it became impossible to confine Wiji in the bathroom since she insisted on being allowed to roam at will. Although she was still fairly weak, she could have covered a long distance in a very short time and we simply could not sit watching her all day. There was the danger from birds of prey who constantly wheeled overhead and would think nothing of carrying off a three-pound mongoose. Wiji was obviously aware of this danger herself, for she turned her pointed face upwards anxiously each time the shadow of wings passed overhead.

Mr Seb, our Alsatian, was put on guard duty, tied to the tree in the centre of the lawn which Wiji loved most. He did keep a watchful eye on her, but mainly, we thought, in self-defence since she had already taken a painful nip at his nose when he had tried to fraternize. The genet, not far away in his day-cage where he now spent his sleeping hours, mewed loudly as soon as he became aware that his playmate, Wiji, was near. We had allowed them to touch noses and greet each other, but the rest would have to wait until Wiji was stronger and completely free of mange.

Finally we hired our gardener's little son to watch over our patient. He was a very small boy of just over seven years old, but he did his duty well. With stick in one hand to protect them both and usually a biscuit or banana which I provided in the other, they would watch each other hour after hour. When it became too hot Wiji moved on to the stone floor of our verandah and cooled herself. Then, after a lap or two of water and perhaps a suck at the egg I had cracked open for her, or the bowl of honey to soothe her throat, she would return to the bare mole-patches and dig, between long catnaps, lying flat on her stomach with front and back legs fully extended.

Once only did our mongoose-guard lose his charge: it was just after lunch and perhaps, in the hot sun, he had dozed off for a few moments. When he came to Wiji had disappeared, though Jessie and Mr Seb still lay in the same place, which meant that no intruder, neither a dog nor cat – nor bird – could have made off with the mongoose.

While the boy frantically searched the garden and verandah and shrubbery, Wiji had already reached the first stage of her exploratory tour, at the end of which she had entered the kitchen at the other end of the house where I happened to be working at the time. Without ado, considering that she had never been there before, she acted rather like a housewife who is at home in any kitchen. She found the Frigidaire first, then went up to the door to make sure it really was closed. After that she wandered near the stove and working area, sniffing at bits and pieces which had fallen to the floor. Then she explored the dog's eating dish, checking its bottom before she took a lick. I looked on, fascinated, and was joined by our maid, Jenny, who exploded with mirth at the antics of this strange little beast which covered every nook and cranny in the space of a very short time.

Eventually Wiji, taking short rests here and there to recover her breath, found her way right round our house, ending up in the bedroom, below Toni's desk, where she laid claim to one of his two very large safari boots. When Toni, whose feet were inside them, tried to move, she objected strongly, squeaking and chirruping in protest as she clung on for dear life. She made a terrible fuss when I confined her back into the bathroom that afternoon though she could hardly keep her eyes open, dropping off to sleep almost at once in her bed of hay. After all, this had been her first day out after a long and critical illness; she was on the way to recovery, but could easily have a relapse.

Unlike her friend Alicat, the genet, she harboured grudges and resented any sort of discipline which had to be imposed on her. Did this mean she was more, or less, intelligent? She was certainly more highly strung, while the genet was easy-going and relaxed by comparison. They were both ready to go home, though with whom and where they would live when Orio and Iain left Africa, we could only guess. Perhaps Daphne Sheldrick, in Tsavo National Park East, who cares for so many orphans, would not mind if two more were added to her number. Or they might return to their camp in Manyara where the new scientist in charge might give them a base they could always come home to. Two animals, semi-tame, could not be cast back into the wild too suddenly, yet to confine them would be depriving them of their birthright.

Would Ali's jaw ever again become completely functional? Would he become strong enough to challege a wild genet and create his own territory? Would he and the mongoose remain friends or would they grow apart, as the genet became more independent? Perhaps, some time in the future, we might, with luck, know some of the answers, although keeping track of wild or even half-wild patients was far from easy.

Chapter Three

LOLITA

At least Lolita, from whom we could hardly bear to part, did not have to leave for an unknown, inaccessible destination. As soon as she no longer needed our care, we decided to accept Daphne Sheldrick's kind offer; that she should join her orphans at Tsavo.

We had named our little duiker Lolita because, Toni said, her eyes were the most limpid and seductive he had ever seen. Needless to say, she was an entrancing female, belonging to the family of the antelope, the hunted ones, ranging from the royal, spiral-horned kudu and eland to the miniature suni.

Because of their behaviour pattern the duiker have survived better than many other kinds of antelope. When alarmed they do not run a little way and then, out of curiosity, stop to look back as do most of the plains antelope, such as the wildebeest. A duiker bounds off, head down, looking neither to right nor to left, taking huge leaps into the air every few paces which must be very difficult for, say, a pursuing caracal-cat to follow, or a farmer, trying to eradicate a crop thief, to shoot. As soon as the duiker reaches cover, it literally dives into the undergrowth (duiker is Afrikaans for diver), but even there it doesn't rest. From then on, until it feels completely safe, it takes a zig-zag course through the bush which usually confounds even the cleverest tracker.

Lolita came into our lives, unannounced and unexpected, having been accidentally found along the Mombasa–Nairobi road by our friend, Doctor Phil Glover, Director of the Tsavo National Park Research Project. She had been tied, four feet together, by a local tribesman who had perhaps hoped to sell her to a passing motorist for a few shillings. Phil, who could not resist stopping when he saw the fawn-grey bundle, paid the price and put it into the back of his Peugeot, hardly knowing what it was he had rescued.

Fortunately, Toni was at home when he arrived and so were Hans and Ute Klingel, staying with us for a few days before laúnching into their new Ethiopian study of the wild ass, a species (like so many others) threatened with extinction. I was on a camping holiday with our son Guy and only due home the day after. However, Ute, who has a particularly wonderful way with animals, had at once unearthed a feeding bottle and teat from a dusty shelf and had somehow managed to persuade the little orphan to take milk the morning after it had arrived. She was only a few months old, weighing a bare seven or eight pounds, standing just twelve inches at the shoulder and in a state of numb shock and collapse, with cuts and bruises on the face and forehead. So Toni had tranquillized her almost immediately and in addition had injected cortisone, the shock-antagonizing substance secreted by the adrenal glands. Had he not done so, she would almost certainly have died of fear, as so many wild animals do when exposed to extreme stress and change of environment. This applies especially to wild birds and antelope who apparently cannot tolerate, either mentally or physically, the traumatic effect of man-handling or excessive unnatural movement.

Instead of causing the usual midnight disruption to our lives, I had managed to book on a morning flight which touched down in Nairobi just after lunch. Toni, as always, was there on the waving base.

'Well, what's new?' I asked him on the drive home, after giving my own news first.

'We have a patient, try and guess what it is.'

'Impossible,' I answered him, 'it could be just about anything from a crow to an elephant.'

'That's right,' he said, enjoying the secret. 'I'll tell you one thing right away. We've never had this kind of a beast as a patient before.'

'Then it isn't an elephant,' I said, eaten up by curiosity. 'Is it sick?'

'Well, just shocked really, and displaced. Phil Glover brought it, and said that if you like, it's yours, even after it recovers.'

When we arrived home, before even the car doors were properly open, Jessie, our jet-black bitch of ignoble birth, and Mr Seb were upon us with much affection. Then Jenny appeared and then the Klingels, whom I had not seen for over a year, but were the same vivacious couple I remembered. Throughout this homecoming celebration Jessie's irrepressible high-pitched howl of gladness rent the air and other than muzzling her there was no way of damping her boundless enthusiasm. To me, it was all marvellous; I was home and the house was full of lovely, sweet-scented flowers.

Toni guided me through the dining-room, on to the verandah and down the garden slope which Ute, complete with baby bottle, had somehow reached before us. She approached the wired-in run and called out to the occupant of the wooden den, built into one end.

'Come on, sweetie, come on; it's bottle time.'

For a moment, nothing stirred and I wondered what would emerge; perhaps a young eland, but the opening was too narrow and too low for that. Another cheetah cub, or lion cub maybe – we had never nursed one before, other than in the wild. Then suddenly, from among the hay and lucerne emerged a graceful, fawn-like creature with the deepest brown eyes and most delicate legs and hooves I had ever seen. She was obviously a little shaky and nervous, but she nevertheless came straight for the bottle which Ute offered her through the wide wire-mesh.

She began to suck, and, as she did, the tuft of hair between her ears danced merrily on her brow. A dark, unbroken line, reaching exactly from tuft to nose, seemed to divide her mobile sensitive face into two clear equal sides. Later, as I grew to know her better, I learnt that her features differed as much between the right and the left side as do our own. Below and just in front of each eye the scent glands were placed in a half-inch dark slit, used for marking territory.

While she sucked at her bottle I stroked her face, discovering two horn buds under her soft fur.

'How female she looks,' said Hans who was sitting back on the lawn like the Lord of creation. 'I wonder if she will stay as quiet as this once the tranquillizer wears off?'

She had finished her bottle and was at the wire fence, trying to nuzzle us, bleating in a friendly, confiding sort of way.

'What an unbelievably beautiful creature, and so tame,' I said. She was the first young duiker we had ever had and seemed to come straight out of a story book I had read a long time ago.

Our first task would be to find out exactly what food she preferred. We guessed her age to be about four months; high time to give up the bottle – which nevertheless linked her to us and made her very tame – and browse full time. (A duiker does not graze.) If she were living naturally her milk supply would have been exhausted by this time. How was it then that she clung to the nipple-image now? Had she been adopted early and, if so, by whom? Surely the man from whom Phil had bought her had not been the one to rear her? We decided that once she was strong enough to go foraging for herself, we would take her into the wildest part of our very wild garden at the end of a harness.

For the next few days it was a matter of trial and error. Ute and Hans had left for Ethiopia and I had returned to my normal working routine. Each day I brought some new plant or vegetable to Lolita, as we now called her, hoping that she would find them delectable. Her wounds were healing and, though she was obviously more secure each day, we could tell from the alarm twitches of her nose and ears that she felt a certain bewilderment. We decided to let her acclimatize naturally since she was doing so well, certain that with gentle constant handling and proper feeding we would be able to win her confidence completely. Our biggest worry concerned our own dogs; we assumed she held them in terror and so tried to prevent them from approaching her run. I had begun to let Lolita out little by little, restrained at the end of a rope which fitted on to a comfortable broad-webbing girth. She loved the change and our company, resenting bitterly when she was, once more, put back into her own domain.

Then our daughter, Gail, arrived for her school holidays and, like all of us, was enchanted with our new patient who had now recovered sufficiently to increase her exercise. For hours at a time she and I and Lolita would explore the steep sloped wild path at the end of the lawn which was covered with wild creepers and shrubs. The little duiker relished these outings, stopping here and there to nibble and to browse; any plant she liked particularly, we would gather in quantity and take back to her run so that she could continue to feed.

But a duiker, it seems, is not conservative in its eating habits. One day she seemed ecstatic about certain berries, and ate them ravenously. On another day she adored the fallen leaves of the gum tree, or the chickweed that grows between the blades of grass on our lawn. Wonderful, we thought, now we have found the answer! Yet the next day she refused them all, even when she was obviously hungry. There was, of course, always the twice-daily half pint of milk and vitamins to assuage her hunger. But since she had well-developed molar teeth she often chewed through the rubber in her enthusiasm to extract the last drop.

'My baby bites through these teats,' I said to the puzzled chemist from whom I had bought a stock. 'I hope she will soon begin to lap.'

Lolita had no difficulty in expressing her desires: she wanted her bottle, her solids, her herbage, our company and the freedom of the house. Since Gail was there to watch over her we began to let her wander about more freely without her harness. The dogs, we thought, could probably be trusted with her, yet we could take no chances without a test. First we allowed Jessie, who was usually very forbearing and patient with all strange animals and birds, to sniff at the duiker while we held her. After all, a dog's natural instinct would surely be directed towards attacking of a prey animal . . . but Jessie, true to her training, just stood, looking longingly at the duiker's bottle, and licking a drop of milk here and there as it escaped down her chin. When we loosened her she sniffed to investigate Lolita more closely, ending up by getting a sharp nudge as the duiker indicated that she thought she had been investigated enough. There had been no drama at all. Now it was Mr Seb's turn.

We decided to wait until the following morning, for two dogs in one day might prove too much.

Mr Seb, sleek, black and very wolf-like, looked much fiercer than he really was. He had been given to us by friends who had returned to America and to whom the idea of destroying such a fine dog was out of the question. We had lost our wonderful old Alsatian dog, Rolf, just a year before and missed him terribly. He had acted as a sort of nurse-cum-watchdog for our wild animal patients and when he died, mainly of old age, there had been a terrible gap in our family circle. Sebastian, whom we renamed Mr Seb, had entered our lives just at a time when we needed him most. Jessie was a good guard dog, but not of a very fierce demeanour. We preferred to have two dogs, for company to each other and to give support when they were on duty.

Mr Seb had been very aggressive when he first came to us; we were warned not to let him anywhere near small children and already knew that he had a way of picking out special people whom he tried to intimidate whenever they came to visit us. But Jessie, who loved people with all her heart providing their motives were honourable, soon taught Seb the error of his ways. Within one month his fierce looks and growls diminished; he became more tolerant and less suspicious and stopped terrorizing nervous guests with his rasping snarl. He obviously had no intention of joining in Jessie's welcome chorus, judging from his bored, slightly disapproving expression, perhaps feeling (as we did) that her deafening serenade was rather hard on the ears.

But within three months he had settled down, establishing his contacts with neighbouring dogs and prospective bitches. He had bullied Jessie from the first, taking the part of the dominant male from the day he entered the house. This surprised us a great deal for we expected Jessie to be top dog in her own territory for a little while, at least.

How little we understood our dog. From the very beginning, Jessie, the extrovert, who rushed fiercely at any strange dog or cat who dared even the precincts of our garden hedge, who would probably have taken on the entire dustman force of Nairobi had they entered through our gate, that same Jessie now submitted meekly to the stranger in our midst, the muscled, sleek, black, fierce-eyed Alsatian.

'Hold her tight.'

I had taken up two holes of Seb's collar and held him securely on the leash. Gail, a little way distant, was feeding Lolita carrots, which she adored. I spoke to Mr Seb, who was sitting up, ears pricked, looking very interested indeed. I studied him carefully; he did not look as if he was going to attack.

'What do you think?'

Gail, who had helped with animal work for years, and who was now in her last year at school, prior to studying Zoology and possibly Ecology later on, already had a very good instinct about animal behaviour.

'I think he'll be fine. Let him come a little closer.'

I moved forward with Mr Seb, until, gradually, he stood almost level with Lolita. She, incredibly calm and unconcerned, moved towards him and began to push at him, the way she always did before taking suck. In the wild this sharp battering of the mother's udder stimulates the flow of milk; here, though under artificial conditions, the duiker still had need of this self-expression which obviously satisfied a psychological need. I must admit that when that little ten-pound body (she had put on two pounds) began to push and nudge the dog my heart sank. He had only to respond with even a mild snap of his huge jaws to inflict terrible damage to the small, fine-boned body.

'Don't *worry*, Mum!'

Gail had seen my apprehensive look; but she was

obviously quite confident that Mr Seb would not harm Lolita. And indeed I need not have worried, for Mr Seb behaved perfectly. By the time Toni dropped in for his mid-morning coffee, the tiny fawn lay chewing the cud next to the big black wolf, both seeming to enjoy each other's unrestricted company. Every time Lolita moved, or flicked her sensitive ears in the direction of a sound, Mr Seb would respond by also pricking his ears. He sat facing her, obviously on guard, repelling Jessie whenever she approached to join the family circle with a nasty earful of growls. He had found a new friend and was not prepared to share her with anyone.

A new era now dawned in Lolita's life: the dog–duiker test having been passed successfully, she was free to come and go in every part of the house, though when she entered the garden watch had to be kept over her, even when tethered by her harness. She did not like to be helped up the slippery stone steps that led to the verandah. Even if her match-stick legs did give way sometimes, sliding four ways at once (we were always afraid she might injure them), she wanted to do it herself and would accept help from no one. Never once did she pass dung pellets in the house. The place for that was in her run and, what was more, a special area of the run. Such incredibly meticulous cleanliness was not, we knew, a feature of duiker and rare among antelope. Lolita just happened that way – and this made her all the more exceptional.

On occasions when we had evening guests crowded into our lounge, Lolita, unperturbed either by cigarette smoke or densely packed humanity, would lie quietly on our thick Swaziland mat next to an enchanted guest and chew the cud philosophically.

'She is like a fairy-tale princess,' one young man said, gazing into her eyes.

'Time for bed.' Gail, noticing that Lolita was beginning to lose her vitality and let her silky eyelids droop, whipped her charge off into the warm, straw-bedded house, content for once to be left on her own.

Towards the end of Gail's holiday we had to make a

decision about Lolita's future. Her head wounds had healed, she had put on weight and several centimetres in height and length and seemed to have recovered completely from her original shocked and concussed condition. Phil Glover, who had kept an eye on his foundling, brought us the offer of a home for her in Tsavo National Park East with Daphne and David Sheldrick, the Warden. We considered carefully what was to be done, for we all loved the little duiker and wanted to keep her. She was safe in her run, but once she ran free there was the ever-present danger of strange dogs. Our garden area was limited and so was our time; we could never really give her the sort of home she deserved.

Two days before we were due to leave for Tsavo, she developed an acute digestive disturbance. She bleated pitifully whenever we approached and when we let her out to pick her own food and to give her suck, she almost keeled over with weakness and with pain. Exactly like a small child, she complained to us audibly whenever she saw us, but calmed down when we comforted her. She had become quite a scavenger, dipping her head into every pot and pan rather like the mongoose. She loved the dogs' Primeal, apples, celery tops, artichoke leaves (but they had to be cooked) and especially she enjoyed home-made bread. Since this was made with whole wheat grains, wheat germ, oil and honey, we suspected that she must have taken what had been put out for the dogs, and, as it was of heavy consistency, had not been able to digest it.

Twenty-four hours after her first attack of colic and with the help of medicines, she passed dung and recovered. Her 'blown-belly' appearance disappeared as if by magic and her appetite returned. We had already considered putting off our journey to Tsavo on account of her illness; now, with great relief we resumed our plans which included taking her for a short trial run in the car to see how she would take it. If she reacted badly, struggled or panicked we would take steps to sedate her for the longer journey.

Gail and I gathered together all Lolita's familiar things:

a box of straw, intermingled with lucerne, her bottle and a spare teat, some corn-cob leaves, young carrots, a bowl of Primeal made up just the way she liked it; and we included one digestive biscuit and some rose leaves as a standby.

Gail got into the back of the car with Lolita and settled her between her lap and the up-ended box of straw, letting her suck the bottle for comfort, after which she explored the car seat from end to end. I started the engine and let it run for a few moments. Then I turned it out of the driveway into the main road. Soon I heard a scuffle behind me and slowed down, to find Gail was having trouble as Lolita lashed out with her knife-sharp hooves: the vibration of the motor had thrown her into a state of terror.

I stopped for a moment to assess the situation. Gail was doing her best to protect Lolita's legs from damage and had wrapped her into a soft blanket. But she bleated and would not be comforted. She would have to be tranquillized for the journey.

'Try and hang on for a bit longer,' I said to Gail over my shoulder as I started the engine. 'We're going home.'

That night we made all the necessary preparations for an early start and discussed sedation. It would have to last at least six hours and if possible eight for, once arrived, it would be best for Lolita to accustom herself to her new surroundings in a calm state of mind.

Toni, as always, was experimenting with safer and better methods of tranquillization. When I questioned him about Lolita he muttered something about trying an entirely new non-chemical method this time; and that I should leave it to him.

I told Gail what was in the air just before we put our heads down that night.

'Leave it to Dad,' she said, 'he knows what he is doing, and *don't* worry!'

Chapter Four

JOURNEY TO TSAVO

How often have I been criticized by my colleagues and family for becoming too emotionally involved with many of my animal patients. A good veterinarian, they say, should maintain a detached interest. But how not to get involved when sympathy with one's patients is very much part of therapy? It is after all invariably a two-way affair. Once the animal trusts you, it gives its affection unconditionally, which usually means that it will tolerate painful handling and manipulation to an amazing extent.

Gail was right once again. Worry never really helps and even total involvement should not imply day and night preoccupation with one's cases. But I need not have worried at all, for Lolita took the two-hundred-mile journey like the most seasoned traveller. Whatever it was that Toni found to calm her worked like a charm and continued in effect for the entire journey and after our arrival when she might have well been disorientated by her new environment.

We rose at five, hoping to avoid the heat of the day and the heavy traffic, estimating that we could reach our destination well before lunch-time. The chin-spot flycatcher was already sending forth his descending, six-note melody when we went down the garden to fetch Lolita. That meant we were late, for he usually began to sing just before dawn; or had the overcast August sky kept him late in his nest too?

Lolita, long since awake, paced her run, bleating softly as we approached, impatient for food. While Gail bottle-fed her I went back to the house to give Toni his breakfast.

While we drank our tea Lolita took her second course of soft crumbled Primeal out of Mr Seb's dish, while he stood by, ears forward, watching her with approval and affection. Only when Gail had led her to the car and lifted her in, did Mr Seb's demeanour change. He knew we were going on a journey and of course he wanted to come. But this time we would be entering a National Park where dogs were not permitted; as soon as he realized he was to be left behind he became dejected and nothing would cheer him until our return home. Jessie accepted our departure activities more philosophically: she would go mad with joy if we let her join us, but was pleased to stay if we didn't. Her ever-cheerful personality would probably, eventually, impress itself on Mr Seb too.

Gail and the little duiker sat on the long back seat side by side, enjoying themselves hugely. Gail had brought all the paraphernalia for our de-luxe passenger: a soft blanket to protect her legs in case she grew restless; lucerne and carrots, milk and so forth. As soon as the car's motor began to hum, Lolita settled down, her fragile-looking, yet very reliable legs bent under her; although she had not eaten much solid food that morning, she nevertheless began to chew the cud in very adult style, rather like a teenager chewing gum to pass away the time. In contrast to her desperate nervousness on the previous day she now seemed to enter into the new experience joyfully.

Our first port of call was the service station where the car had to be filled up. Lolita stood up for the occasion, watching the activity with intense interest, like any young passenger to whom car-travel was a novelty. The garage attendants were fascinated. One thought at first that she was some sort of dog, otherwise how could she behave like one? But the windscreen cleaner, leaning in at the window, soon put the other man right. 'Swara,' he said with assurance – which means: antelope. Others crowded round and wondered. Not, we thought, because they had never seen an antelope before, but because perhaps they had never seen one that was car-trained and completely tame.

A few minutes more and we began to leave the city environs, reaching the first open country still containing game. First the open ranchland, dotted with tall acacia and ground-level bushes of whistling thorn, relished by the giraffe which we saw in groups of five and ten against the slightly-shrouded sunrise skyline. Gazelle, revealed only by their white twitching tails, were dotted all over the grassland which rolled towards mountain country on each side: the Ngong to the south-west and Mount Kenya to the north, still wrapped in heavy cloud which in the rainy reason might lift just before sunset.

More wild animals were plentiful, even on farmland. Gazelle, Grant and Tommy, warthog, cheetah and zebra, though the last more rare. But they were there and so were lion and leopard, wild dog and hyaena; and so was the farmer, with his gun, protecting his stock. Eighty miles down the road, near Sultan Hamud, we passed Norman Peckover's 10,000-acre ranch where a herd of eland was grazing the roadside pasture. Norman prizes his wild grazers and browsers; he protects them, even nurses them. Since he is a progressive farmer he knows that wild stock and domestic cattle entirely complement each other. Vegetationally, they do not overlap; socially they have no quarrel. The eland browses down the sodom-apple which is deadly for the cattle; as a result of this it doesn't have much chance to re-seed. Giraffe browse the tops of the taller bush until much of it is destroyed. Less bush means less tsetse-fly and hence less of the dreaded sleeping sickness which kills domestic stock. Norman counts reedbuck, gazelle, waterbuck, zebra and some of the smaller antelope among his resident wildlife population, of which only zebra and impala compete with the cattle for grazing.* But his losses to predators are heavy: 117 of his stock (wild and domestic) in four years.

The beautiful, deadly predators; would Lolita be able to survive living in a natural Park area without falling prey to a wild-cat? Did she know the danger instinctively? Did she know the sounds which bring quick death? She had not been afraid of our dogs when she should have been, as Alicat, the genet was from the first, his body and tail crest rising whenever he heard the faintest canine sound. Wiji, the mongoose, loved dogs for reasons best known to

* *Africana*, Vol. 4, No. 1, Richard N. Denny.

herself. Was fear, then, inherited in some animals and not in others? Should we have left Lolita safe but restricted in our town enclosure? I thought of Daphne and David, and their years of experience with orphans; if anyone could protect Lolita, yet at the same time give her independence until she might, one day, wish to leave because she felt safe in the wilderness, then it was the Sheldricks of Tsavo.

From Mtito Andei on, which lies one hundred and fifty miles east on the tarmac road from Nairobi, Tsavo National Park, Kenya's largest Wildlife refuge, stretches both to right and to left. On the right and south lies Tsavo Park West, a spectacular incredibly diverse wildlife sanctuary, containing the world-famous Mzima Springs whose waters rise in the undulating Chyulu Hills to the north-west. Tsavo Park East, lower lying than Tsavo West and thus ecologically different in some areas, is the larger of the two and contains enormous tracts not yet developed for visitors. We could see the distant Yatta Plateau, vast and almost unexplored, hazy and remote in the mid-morning heat. Whatever the time of day, leaving the greener flowing hill-country of the uplands and entering the mysterious stark wilderness of Tsavo was always exciting. This was the trackless, hot scrub country the pioneers had traversed and conquered, yet here we were, skating along at great speed with no thought, except to reach our destination!

Clouds gathered in the vivid blue sky, driven by cross-winds which one suddenly encounters in the channel between the mountains; there was promise of rain in the air which would tinge the brown-yellow rusts with green if only the wind would drop.

Tsavo East had once been in jeopardy because of drought; some had predicted that it would fare worse, very soon, and be transformed into desert. The elephants had been held responsible for the destruction of vegetation which other species needed, yet without them many animals who depend on the water-holes that they have dug would have perished. Five thousand had been threatened with destruction, in the name of science. The Tsavo controversy had raged not only in Kenya but in many parts of the world and for a time things looked very black for the elephants, of which three hundred had already been slaughtered.

David Sheldrick, the man directly responsible for the creation and maintenance of this vast wildlife stronghold, decried the slaughter of the elephants which seemed not only premature but ill-advised. Many views were voiced, among them Toni's and other eminent scientists, the burning question being: was the future of the Park really endangered by its twenty thousand elephants? No one could look into the future with certainty, nor was there a precedent from which to compare the present situation.

'When in doubt, don't,' goes the saying; and so, at the eleventh hour, almost by miracle, reason prevailed and the slaughter was halted. Dr Glover took over the reins of the research centre which would concentrate on ecological studies involving not only the much-maligned elephant but the vegetation and the inter-relationship of all populations involved.

Phil, a gentleman and brilliant scientist who loved and respected the creatures he worked amongst, soon confirmed his own and David Sheldrick's prognostication and hope concerning the Park habitat: the flora were undergoing a natural cycle which was bringing a new look to a large area of the Park. The pattern was changing before their very eyes; the elephants, destroying Commiphora and other trees wherever they moved, had opened up dense shrub and tree stands, thus encouraging grassland and increasing the carrying capacity of game enormously. More plains animals, such as oryx, zebra, eland and buffalo appeared in larger numbers; and browsers such as gerenuk, lesser kudu and black rhinoceros were seen more frequently and were living off shorter bush which had become more plentiful. Aerial counts showed that the elephant population was *not* increasing, as had been predicted. There was, in fact, no change in their number since the count three and a half years before. Were the elephants limiting their birthrate because they were overcrowded or were they migrating to outlying areas to relieve the pressure?

Toni came to a sudden halt; I had been watching the drifting cloud shadows on the angular rock horizon and had not seen the huge ears and trunks at the side of the road.

'Good heavens, they are almost pink!' Gail exclaimed from behind me.

I looked back and saw that Lolita, obviously taking a lively interest, was standing up and seemed to be gazing at the muddied giants attentively but without fear. They were two young males, who, between mock charges, were chewing bark from a prehistoric-looking baobab tree whose very soul seemed as elephantine as the two giants who fed from its trunk. This was the season of the first long-awaited rains; the elephants turned from dignified slate-grey to the red colour of the sticky mud they wallowed in. We watched until, replete, they wandered off into the dense scrub of the sanctuary. The moment they entered it they became invisible, in spite of their rufous hue which somehow harmonized incredibly with their surroundings.

It was just about midday when we climbed the last hill to the Sheldricks' home. We parked our station-wagon next to the Bougainvillæa bush of white and pink which always delighted us anew. Daphne Sheldrick and her two daughters Jill and Pippa were there to greet us. Wiffles, the dik-dik, was there too, browsing in the exotic, colourful garden which Daphne and David generously shared with all their orphans.

Gail lifted Lolita out of the car and put her down in the driveway. We stood there, all of us, wondering what her reaction would be, but immensely relieved when she remained serene. Without taking any notice of the dik-dik who eyed her inquisitively, but kept his distance, Lolita began to wander on to the lawn in search of food. We trailed behind her, apprehensive in case she would take fright and career off at top speed.

Daphne had an enclosure at the side of the house and to this she now led us. It would be a good acclimatization pen for Lolita until she got her bearings. It was roofed in with wire netting to keep out predators, such as leopard and wild-cats, and strong enough to prevent hyaena from chewing their way through. We entered and Lolita followed without question; then she stopped, looking slightly disconcerted for the first time since we had left Nairobi. She was looking intently at the rabbits, animals she had, perhaps, never seen before in her life. They stared back at her, then continued to nibble at their food. Lolita slowly walked towards them and bent down her head to their level. Then in the manner which is so special for wild creatures belonging to the same ecological niche, they began to commune, mutely, wrinkling their faces and touching noses as if to make each other's acquaintance. The ceremony complete, Lolita relaxed and inspected the length of the run while Wiffles, who had followed behind, observed her rather coolly from a distance.

'Time for the bottle.'

Gail had prepared the duiker milk and was passing on the routine to Daphne who had years of experience of raising animals of every kind. I continually urged her on to write another book to follow *The Orphans of Tsavo*, for her knowledge of animals was immense.

'One is quite enough,' was all she would comment; but I had the feeling just the same that it would not be long before she would be back at her typewriter.

Meantime Lolita, very thirsty from her long drive, sucked eagerly, chewing through her third teat that week. When she had finished she lay down near her new-found friends to take her customary midday rest, every now and then stretching her long neck to take a chew at their food.

We left her there and drove to announce ourselves to our hosts, the Glovers; parting from Lolita the next day wasn't going to be easy for any of us.

That afternoon a young Dutch ornithologist, who had joined the Tsavo research centre, took us on a four-hour drive to Lugard's Falls. He had only been in Africa for two months, yet his knowledge of animals, and especially anything on wings, was already tremendous. He did not only try – and succeed – to recognize each and every nest as well as its occupant along the way but each pair of wings, as they flitted past. From the woolly-necked stork to the sand plover, the doves, the rollers and hornbills, the woodpeckers, larks, shrikes and weavers, we identified them, one and all.

From her perch on the open roof-hatch Gail had a bird's eye view of everything. A huge herd of black-brown buffalo, their dung still fresh on the road, stood solid and determined, challenging us with horns proud and high. Along the river, shaded by palm thickets, family groups of elephants were bathing and playing along the banks. As we approached some clambered up the steep sides in alarm, while others, the elder matrons, either continued their ablutions or mock-charged us, ears flared and trunks up. They knew and we knew that no one was going to be disturbed; their display was a matter of form to preserve their dignity and ensure our respect. Before we had reached the edge of the acacia woodland which thins near the rock-faced arena of the waterfall, we had seen lesser kudu, gerenuk, eland, zebra, waterbuck and black (hook-lipped) rhinoceros, all in a matter of one hour and a half. We had seen no predators, except those on wings like the tawny eagles and pale-chanting goshawk, hovering as they hunted for prey.

A tourist had, very recently, fallen into the river just above the waterfall. He had wanted nothing more than to photograph the rocky gorge with its fantastically carved rocks, and to do so he had straddled a bridge of stones. He had been swept down through the tumultuous narrow waterfall before anyone could help him. At the bottom of the rushing, swirling water, the River Galana widened into a pool. With our binoculars we counted eleven enormous crocodiles sunning themselves on a rock, an excellent vantage point for spotting prey that was being swept towards them. They looked terrifyingly saurian as they lay there, still as statues in the sun, some with their jaws open, their armour-plated forms camouflaged amazingly into the grey-brown smoothness of the sun-baked stone.

Only those who live year in and year out in the wilderness, whose perceptions are not weakened and numbed by the abuses of modern life, know what is out there in the wild thickets and woodlands and rockshelves we had passed that afternoon. Oh, if only we could remain in the wilderness longer to regain our bush-sense and calm our city-frayed nerves.

Yet once again, our stay in Tsavo Park had to be short. We returned to the Sheldricks' home on the next morning to take our leave of Lolita, who was obviously revelling in her freer, more natural surroundings. She had settled in well. According to the Sheldricks she had walked into the house that evening and explored it exactly as she had done in ours. She had then taken possession of the dik-dik's food and devoured it. Wiffles had watched without protest; he was either very courteous or very frightened or, like Mr Seb, had not been able to resist her enormous sex-appeal.

Wiffles, whom Daphne had raised from a week-old fawn, was quite invisible as we entered the garden, blending completely into the luxurious flowering shrubs. With a lightning movement of his head he looked up as we approached, his mobile elongated nose directed at us as he gave a high quivering whistle with nostrils distended and his red hair-crest raised. Then, for a moment, he stood regarding us with speckled gold-brown eyes, scenting the drift of soft wind that moved up from the valley. The rounded scent-glands had the effect of accenting the brightness of his puckish, alert features, in complete contrast to Lolita's gentle, calm, feminine personality. Wiffles had long outgrown the bottle and had already been on forages in the Park, seeking a mate. Since dik-dik live in pairs, perhaps Wiffles would soon be bringing home a wife!

Lolita came over and touched us, each in turn, with her soft damp muzzle. Then she nudged Gail, looked up and bleated softly, as if trying to tell her something very important.

'She wouldn't take her bottle from me this morning,' Daphne said, 'and I am *sure* she is thirsty.'

Was it scent, or appearance or manner, which had caused this block? Duiker can go without fluids for long periods, when necessary, but will drink when water is available. Lolita, used to her pint a day, could not be deprived of the bottle so suddenly; unless it was tailed off gradually it might well cause a severe psychological trauma, just as it would in a suckling baby.

Gail took the bottle, already prepared, and held it out to Lolita. At once she made a rush at it, pushing and pulling at the same time, almost falling over herself to get the teat. With eyes shut in ecstasy, her long, well-muscled neck stretched up, she drank or rather gulped the milk, to the accompaniment of rhythmic gurgles. Daphne took the bottle from Gail very gently and we thought Lolita had not noticed but she lifted her head, opened her eyes and saw the new face and at once stopped drinking, retreating back towards Gail. There was nothing for it but to begin again; this time all went well. When Daphne took over, Lolita checked only for a moment, then continued until the last drop had gone.

The last time we glimpsed her, she stood at the top of the steps which led down into the Park wilderness. She stood quite still, framed against the purple hills, a little nervous, we thought, a little reflective, as if aware that she hovered at the edge of freedom which was so unbelievably vast that the enormity of it overwhelmed her.

Chapter Five

THE RESCUE – AN INTERLUDE

My sister, Alice, would have loved the gentle, Bambi-like Lolita; she had been born wild and yet, in some inexplicable way, had shown complete trust of everyone, even our 'predatory' dogs. She seemed like a symbol of all that is non-violent, her fearlessness endowing her with a power to keep everything fearful at bay. Nature at its gentlest, is what Alice wished to portray. 'I'm so looking forward to coming to Kenya,' she wrote, 'but please, do remember that I am a city girl and that I don't want to go on any wild safaris – working or otherwise – with you.'

We had, for years, faithfully reported to her in our weekly letters our journeys into the wild. Perhaps our stories of encounters with elephant and rhino and the larger cats loomed large in her memory now that she was about to launch on her journey. She had written that she would like to see as many of the large wild animals as possible and, if not to see them, hear them; and, if failing that, to find some evidence of them, even if it was only their dung.

'Steaming dung of elephant is almost as good as seeing the animals,' she used to say. 'Then I can tell my friends that I passed along the road just behind them.'

'Two weeks is all I can spare,' she had written just before she came, 'and I am terribly excited. If I can only sit under some large shady tree and work at my woodcuts, for some of the time at least, I shall be completely happy.'

But would she be? Alice was not only frightened of the large wild animals but the small as well and she would be less than pleased by the kind of companion who might be sharing her tree: a scorpion, or snake, or a very large hairy (but quite innocuous) spider, or a buzzing rhinoceros beetle which might well appear and disrupt her serenity. She had always abhorred frogs and sticky-trailed snails and 'thousand-legged' monsters that catapult into a curl at the slightest touch; and insects armed with a shiny carapace and long waving antennae and others which, still as statues that emulate leaves or sticks, flick into life when one least expected it.

Thus it had always been, from our earliest childhood: Alice, gentle, remote, yet at the same time intensely alive and creative, able to recreate the loveliness of Nature: sensing, rather than experiencing, the subtleties of light and movement. The most commonplace familiar scene, such as a wind-bent glen of tall gum trees along a country road would come to life under her hand, tinted with a special kind of nostalgia as if the artist regretted ever having to leave them. Always they were alive and vital, the submissive tree-crowns melting into the movement of wind-driven cloud.

My own passion, which had always been animals, she did not share in the least, except in a respectful, distant sort of way. She admired animal beauty but did not want to get too closely entangled.

By the time Alice arrived we had formed a plan: Lake Naivasha, with its multitude of birds, its shores of spreading acacia tree-crowns and tall reeds, the restless water of the lake set against a backdrop of stark, newly eroded volcanic mountains and contrasting hillsides and to the north-east the towering Aberdares, all this would provide just the kind of material that she was seeking. If the hippos and insects could keep away and the weather was good, location Naivasha should be an artist's haven.

Alice duly arrived, with two children in tow, amazed that in spite of newspaper reports of leopard visits in suburban gardens, Masai tribesmen's cattle raids on near-city farms and other alarming yet journalistically highlighted occurrences in the Kenya wilderness, the airport and the city was not only attractively planned, but colourful, highly organized and unexpectedly civilized. Even so she did not quite trust outward appearances. Delighted with our wild, aloe-dotted garden and the rocky untamed slope leading to the river below, she could not believe that one could go so far as to take a walk in complete safety.

'What have you got hidden away down here?' she asked a little nervously as we tried to persuade her to join us for a walk to the river where the dogs loved to swim.

'We have a private python, who keeps to himself,' I said flippantly and immediately wished I could unsay it. But just then two weavers settled on our feeding platform and took Alice's mind off the subject of snakes, for the time being anyway. By the time we reached the Naivasha farmhouse loaned to us by friends, she had acclimatized considerably, revelling in the exotic atmosphere of Nairobi with its diverse society and aromatic coffee houses – and its cosmopolitan atmosphere which completely captured her heart.

Our Lake holiday was an immense success and whetted Alice's appetite to see more. We heard hippo honking and saw their new tracks and the vegetation which they had sprayed with dung to mark their territory; we watched a grey statuesque Goliath Heron and the snake-like darters and the pelicans that haunted the jetty-like papyrus at the end of the sloping lawn which led from the stone house to the water's edge. We sat entranced at the aquabatics of the many species of duck and waders and the acrobatics of the coots treading water while they fed between the pink and purple water-lilies. For hours we strained our necks to see the eagles hunting overhead, foremost among them the fish-eagle, whose clarion call echoed back and forth across the water from dawn to dusk.

The only creatures which disturbed Alice were the humped, Boran bulls who came to share our lush green lawn; they looked dangerous by virtue of their enormous horns and huge bulk even if they were not; being neither tame nor wild it was hard to assess their temperament.

By the time we returned home from the lake, Alice was ready for the next adventure, which, this time, would have to provide more than just sounds of hippo and birds. We cudgelled our brains as to how we could take her among big game without the remotest danger of nerve-racking (for her) encounters, knowing full well that even one mock charge by buffalo, rhino, lion or elephant might dampen her present enthusiasm for ever.

Then we thought of Treetops, the hotel on stilts in the Aberdare National Park where one can view wild animals from above in complete safety and extreme comfort. It was the luck of the draw whether we would see very much of elephant, rhino, or buffalo. Luckily April was the period of the short rains, when the biting fly drives the animals out of the forest into the open.

The Outspan Hotel in Nyeri, which lies seventy miles north of Nairobi on a tarmac road, is the receiving depot for Treetops' travellers; from there they are taken to the edge of the forest by Land-Rover and thence the journey is continued on foot.

We reached the parking area at the edge of the forest in the drizzling rain just before four o'clock in the afternoon. I silently prayed that on this day the larger inhabitants would not arrive at the dam before we did and that our walk to the Treetops hotel would be free of incident. We reached the end of the trail safely and climbed up the spiral steps into the hotel and up to the rooftop, to take tea among the irrepressible baboons.

The vista which spread out before us was superb: thick indigenous forest spread out in every direction surrounding a large clearing which contained a mud-banked dam. Warthog had ventured to the edge of the water: one very grey, warty male with large tusks, one female and three young, tails curled up as they drank. I had so often seen the same spectacle at water-holes: the prey animals hovering in the background, uncertain whether it was safe to expose themselves out in the open, vulnerable as at no other time when their heads were down quenching their thirst. Usually one could see them very clearly; here they were hidden in the thick tree-cover.

Like the regimental standard-bearer, the warthog would appear first, standing to frozen attention for a few moments to ascertain scent and sound. If all was as it should be he would then trot straight down to the water without looking to right or to left, followed by his faithful family, also at a trot, tails up. After drinking their fill they would remain on the banks feeding or wallowing in the mud, perhaps still on guard while the other, more nervous animals came down to take their share of water.

When the warthog family disappeared the red-bodied white-maned bushpig arrived, not to drink but to root in the soft mud, digging with his tusks. Then seven shaggy, long-haired waterbuck filed down the slippery, well-trodden path: one magnificently-horned male and six females. They looked very much more in place in the grey subdued tints of the wet evening than in the flaring African sun. Near by, with the aid of binoculars, we watched a bush-buck doe suckling her young at the edge of the forest. Through the drizzle the last rays of the evening sun made her white-striped body glint red gold.

As soon as the light grew dim a herd of buffalo arrived, giving voice as they settled down to drink and to take the salted mud. They slipped and skidded, often immersed half-way up their legs, yet always managing to extricate themselves. They bellowed and grunted softly as they dug, as if commenting to each other. As the waterbuck slowly filtered back into the forest a kudu bull appeared, hesitating in the clearing near the bushpig. Suddenly the floodlights lit up the arena, momentarily startling the bushpig and the kudu who disappeared out of sight.

As evening deepened we very reluctantly left the chilly rooftop, and the spectacular 360° view into the Aberdares. The floodlit clearing now gave the forest an even denser, more mystical appearance than before, and as we began our descent we could just see elephantine forms emerge from between the trees, slowly and silently, quite unlike any of the other creatures. By the time we had found our comfortable seats on the top-but-one deck they had begun to drink and to dig for salt with their tusks.

There was one breeding herd, mothers with infants, and a separate group of new arrivals, probably all males who drank from the opposite side of the dam. They had posted their sentry; a long-tusked elder who stood alone, lifting his trunk at frequent intervals to test the wind. Once, very suddenly, he gave a shrill trumpet as if answering a summons; at his call the herd of males immediately left the dam and returned to the forest.

The stage below us, more brightly lit as the last of the day disappeared, began to take on an unreal, dreamlike quality. It was raining more heavily and the ground became increasingly soggy. More buffalo arrived, intermingling with the other herd; and then more elephant, a small group of young giants who tried to enter what seemed to be the ring of a family gathering. But the formation did not give way – not at first; then an old female slowly rose from her knees, turned and touched one of the intruders with her trunk: first his head, then his back, gently but thoroughly. As one of the circle trumpeted softly, the others seemed to increase their rumbling, a sound not made with their stomachs, which hold a huge quantity of fodder, but with their throats. To me rumbling had always meant an expression of comfort and trust; they rumbled when they fed, or met together, or when the mother was reunited with her infant. Sometimes, when watching a large herd, one can pick up different tones of rumbles and all, no doubt, have their meaning. But at least, though we know so little of elephant communication, it is certain that the deep 'belly' rumble is not a sound of aggression.

The ceremony of touching the leader of the new small herd took over ten minutes. Several others joined in, as if to ascertain whether or not the new arrivals were worthy to join them. Then, as if by a given signal, two of the circle parted, admitting the rest. Within moments they had intermingled and life continued as before.

We could hardly bear to tear ourselves away from the spectacle. Dinner was being served on the deck above and the dining-room seats did not give an easy view of the dam. But being frail mortals, aware of the rumbling of our own stomachs, we repaired upstairs to eat as quickly as we could.

Just as we were getting up we heard someone say that

an elephant was stuck. We rushed out to find a commotion on the left shore. Gradually our eyes made out a homogeneous mass of churning, dark brown earth which contained an elephant baby, rumbling – and this time there was no mistaking its meaning – and trumpeting in a high-pitched panicky tone which summoned some of the matriarchs. At the alarm two females who had been disporting themselves in the much-coveted salt-lick, separated from the herd directly below us. They made straight for the stranded infant, one of them – we thought the mother – trumpeting as if to console it. The second elephant followed in the footsteps of the first and on reaching the calf took up her position behind it. Bending her front leg against its buttocks, she began to push while the other female, her trunk now entwined round the infant's neck, levered and pulled slowly but gently. Within moments the rescue operation had been completed and all three returned to the previously-occupied place on the shore of the dam.

'Fantastic,' Alice said, 'quite extraordinary. No one will believe it when I tell them. How can one ever doubt the elephant's intelligence after that?'

We wrapped ourselves in warm blankets and again settled down in the easy chairs on the second deck. The buffalo, after three hours of mud-revels, left one by one, disappearing into the forest behind. The front entrance – the opening of the forest on to which we faced – seemed to be reserved for the elephant that night. More came and some left, until we had counted seventy in all and over thirty buffalo. It had all been very peaceful; there had been no disruption of harmony or sounds of conflict. The little elephant was back with its mother who seemed to keep closer watch. There was little sound except for an occasional conversational trumpeting and scuffling and chewing. There was, of course, the background chorus of the insects, interspersed with some gay, very bird-like frog chirruping. We seemed to be gazing through a window of Time, permitted to take a glimpse into the lives of the inhabitants of the forest; as they have been for thousands of years and as we hope they will continue to be long after we have gone.

Suddenly the peace was shattered by the entrance of a black rhinoceros, puffing and blowing as if everything depended on the impression she made. The elephant appeared amused by the commotion, breaking off for a few moments in mock retreats and charges, trumpeting and shaking their heads as the rhino tried to usurp their place. This kind of behaviour, Ruth Woodley (wife of the Warden of the Aberdare National Park) told us, was a regular occurrence, each species testing anew, yet already knowing the strength of the other. Even so the animals behaved as if this was something entirely new and the elephant humoured the aggressive, short-sighted rhino as if they enjoyed the pastime. We watched one long-horned rhino back a young male elephant seemingly helpless and defeated almost against the water's edge. Just as we were about to sympathize with his predicament he stopped dead in his tracks, fanned out his ears, lifted his trunk and pitched forward towards his unfortunate pursuer, trumpeting most threateningly. The rhino, taken aback, slipped sideways to avoid the elephant, then got up and slithered away as fast as he could. The elephant, his little joke complete, then shook his head several times as if to assert himself more completely, after which he rejoined his group.

There were now eight rhino in all, two with small calves at foot. They brawled with each other and with the elephants, as nervous and as aggressive as the giants were peaceful and calm.

'Just as I was beginning to feel I was wrong to think that the wilderness is all strife and tension,' Alice said reflectively, 'these monsters come along and spoil it all. What is it that makes them so quarrelsome?'

But no one could answer that, not even Ruth with years of experience behind her. It seemed just a little unfair that one kind of creature, like the elephant, had been blessed with so much nobility and wisdom, while the black rhinoceros seemed eternally short-tempered and crusty.

It was getting very late; most people went off to bed, but neither Alice nor I could tear ourselves away from the drama taking place before us. We had been assigned the central room which faced directly over the dam and we would be able to look straight down on to it from our beds.

'Let's stay a little longer,' Alice suggested, hardly able to keep her eyes open. The city girl who abhorred life in the raw had been caught up by the excitement of seeing wild animals in the wild.

Just then a female rhino and her calf waded in, knee deep, near the spot where the elephants had effected their remarkable rescue. As we watched them slip about at the water's edge we saw, to our horror, that the rhino calf had become deeply immersed, crying plaintively as the mother tried to locate it. As soon as she approached she was distracted by the arrival of another female rhino who had been drawn by the cries of distress. At once a new battle ensued, accompanied by roars – which I had never heard before – and the clash of horn striking on horn. As soon as the mother had achieved her defence she resumed her search for the calf, a process both unsuccessful and hopeless. Was her hearing as poor as her sight? She seemed to come so close and then pass by within feet, losing track of the object of her mission. By three o'clock in the morning the calf's cries had become feeble and so had our perseverance; we lay, face down on our bunks, following the harrowing drama from our rooms, desperate for the calf in distress and for the mother whose search became more and more erratic. By four o'clock the elephant and rhino had departed; all but the mother and the little one who had ceased to make any sound or to struggle in its death-trap of mud. Just after half past four only the small rhino remained: the mother, having taken water and salt, had sauntered off into the forest to be seen no more that day.

At seven o'clock the Aberdare Park Warden, Bill Woodley, arrived with ropes and a winch to pull out the orphan, but by this time we had returned to the Outspan Hotel.

Ruth Woodley told us that the calf, much stronger than had been expected after its ordeal, charged its rescuers the moment its feet touched firm ground. It was useless to return it to its mother; this had been tried in other cases but had always failed. A rhino female, once the calf is out of sight, seems to forget that she has ever borne one.

We were thoughtful as we ate our breakfast that morning; thoughtful and weary as if we had lived through a lifetime of experience in one single extraordinary night.

'What do you think now?' I asked Alice, confident that the challenge and immensity of the wilderness must have moved her deeply.

'At last,' I thought to myself, 'she will also understand why we fight so hard to preserve the wild and why we do so many hare-brained things towards that end.'

But Alice, overcome with fatigue and emotion, could not see it that way at all. With an intensity and fervid conviction I had only seen her express previously when she gave judgement on some work of art, she declared that now she was sure wild animals were, after all, very much safer in the Zoo.

Chapter Six

VALLEY OF THE CHIMPS

To our surprise and intense relief Gilka did not react when the cage door was snapped shut behind her. It was just past seven o'clock in the morning, an hour after dawn; the time when the chimpanzees began to climb down from their night nests. Someone came over to tell us that a small family group had already been spotted over the hill, heading our way. We would have to be quick if the first anaesthetic injection, which Toni had already prepared, was to take effect before the troop arrived.

Toni dashed behind the palm screen enclosure erected round the cage and pushed the needle home. As he did so Gilka shrieked and screamed in anger and in fear, rattling her cage with tremendous strength and persistence, the sound carrying across the valley and probably far beyond the hill crests. There was nothing for it but to wait and to hope that gradually, as the drug took effect, she would quieten down.

Luckily for us there was a small but very robust hut near by; when the forest above us suddenly became alive with the vocal response of two large male chimpanzees, Hugo van Lawick, in charge of operations, gently but firmly suggested that we take cover.

'If they catch us near Gilka while she is trying to get out there might be trouble,' he explained, but we needed no urging. Having never seen chimpanzees in the wild before we were suddenly aware of their immense strength and irrepressible energy as they came crashing down the slope, whooping as they went in answer to the distress calls which continued louder and more shrilly at every moment.

Gilka, a nine-year-old adolescent female, had been lured into the cage by devious means, but for her own good. She had developed a large deforming growth on her nose and eye-ridges over the past year and had begun to avoid her companions as a result. It must have been irritating, to say the least, but probably also very painful. If it continued to grow it might constrict the nasal passages; if it ulcerated or became an open wound, septicaemia might result. Something had to be done, and quickly; the fact that a wild chimp had never before been immobilized posed a problem, but one no greater than other 'firsts' Toni had faced in the past.

Jane Goodall of chimp research fame – now Jane van Lawick-Goodall since she had married the renowned photographer – had come to see us just before she left for England to launch their new book *The Innocent Killers*. She had brought photographs of Gilka and had told us of her fears: it was suspected that the chimp suffered from Yaws, a disease which showed similar lesions in humans.

Chimpanzees are very susceptible to infection; poliomyelitis and influenza had already claimed a number of lives among the troop of 35 to which Jane had devoted years of study. Though they lived in the remotest possible place on the shores of Lake Tanganyika, the local fishing populations with whom they are in contact, mainly through flies, could easily be a source of disease. A fast-spreading organism to which the chimps would have no resistance could wreak havoc; it could, in fact, wipe out whole populations. Nothing would be sadder than to see the last few nuclei of these highly-evolved homo-linked creatures threatened or eliminated; already there were so few left in Africa. As man encroached and forests disappeared, the battle to protect them would become a full-time occupation for conservationists. At last the species had been placed on the list of animals threatened with extinction.

But now it was Gilka, a female adolescent, who was in danger. She had been a lonely youngster since her mother, unsociable and timid, had kept herself and her siblings separate from the troop. Always on the edge of chimp society, Gilka had taken a baboon as friend and for a year the strange – though not unique – companionship had thrived. Then polio had struck; some had died, others were maimed. Gilka had survived but her right hand and wrist had been affected, remaining slightly weak and shrunken as a result.

We were delighted to know that so much trouble was being taken over one single animal. I looked at Jane whom I had only briefly met before but about whom I had heard so much already: she had braved the wilds years back to study the chimps of Gombe Stream Reserve, as it was then, in Tanzania. Alone at first, then only with her mother for company, she lived a lonely and simple life in the tropical, remote mountain country. The knowledge of chimpanzee behaviour she had acquired during those years was immense, but so was the challenge of trying to understand the many facets of their intricate lives. They were being studied not only as animals in their own habitat, but in relation to their similarity to man. Here was much interest not only for biologists, but for anthropologists, ethologists and for any scientist studying comparative behaviour. Eventually a small group of students had come to Gombe Stream which had developed into an active, very productive research centre. It was hard to realize how many different aspects of chimp behaviour had to be studied separately, or how detailed and exhaustive they were. The troop of 35, not a close-knit society but a collection of socially associated groups, commanded an area of 35 square miles of sanctuary. Beyond Gombe Stream there lived other chimp groups, the estimated total population being between 100 and 150. It had taken years of incredible patience and application to reach the state of trust between man and animal that Jane Goodall had achieved. At first, they had only allowed her to approach, later they lost their fear of groups of people and began to call at the research centre for bananas they would occasionally receive. If they had lost their fear they certainly did not lose their respect; except for one male who frequently bumped or knocked against people in sheer exuberance, they kept their distance and much of their natural reserve while the behaviourists took heed never to become complacent in their presence.

'When can you come?' Jane had asked. Their private pilot would fly us there together with the local hospital surgeon, Dr Roy, who was particularly interested in the chimp's condition and would take a biopsy. Hugo van Lawick, then in England, would return in time to fly us back. Meanwhile we would send tranquillizers for Gilka hoping that she would accept them in bananas; the dosage would have to be very carefully measured, since an inebriated chimp might well injure itself when climbing or descending trees. The object of the operation was to calm her just sufficiently so that she would walk in and out of the cage which had been assembled in her territory near the living huts; and so that, on the appointed day, she would allow herself to be shut in.

We landed at Ujiji on the eastern shore of Lake Tanganyika, where Livingstone's travels finally came to an end, and drove to Kigoma, going up the lake by boat. The two-hour trip had revived us; the cool spray and the fading sunlight sky over the distant Congo mountains was like a fairy land, to the north we could see Burundi, and along the east, the shore we followed, steep hills of tropical vegetation, at the base of which lay the fishing villages of the Ha tribe.

Though nothing spectacular had occurred during the weeks that she had eaten the spiked bananas, she did, according to a report we received, seem very much more calm and sociable. Almost daily she walked into the cage where she was fed, a little timid about the strange new edifice on her hunting ground, but not sufficiently worried to be able to resist the prize that had been placed inside. Toni prepared some phencyclidine in his syringe to augment the narcotic already injected, ready to go out to Gilka the moment Hugo gave the all clear. It was a matter of judging her weight and her tolerance; the first was estimated (certainly underestimated) at 44lb., the second was unknown. It was, as so often in wildlife work, a case of trial and error and, above all, of taking it slowly.

The males, after giving the captive some comfort, had

climbed one of the tall nearby palms seeming uncertain what to do next. Lori, one of the young scientists, made up their minds for them: with a bucket of bananas she lured them away to a remoter part of the hillside, at which moment Toni and Hugo again hurried inside the leaf enclosure. Gilka, frightened lest she receive another prick, came for Toni, hurling abuse and anger at her well-meaning tormentor. This is just what he had hoped for: as she advanced unsteadily he took aim and squirted the syringe contents into her mouth, after which he and Hugo once more retreated to the hut which we had meanwhile arranged as a make-shift operating theatre. Again we waited while our patient rattled the sides with enormous strength and determination. Once more her would-be rescuers appeared, this time a different group including females, but they made no fuss after an initial investigation. I was most surprised that they did not attempt to break down the cage, or at least remain glued to the site. Perhaps Gilka, already different from the rest because of her very strange facial distortion, held no real interest for them. Or perhaps they distrusted the strange unnatural aura of it all, preferring not to be involved. It was just nine o'clock. The freshness of the morning had disappeared; it was going to be a very hot day.

Gilka gave no more distress calls; Toni trajected a little more of the solution into her mouth until slowly she passed into a state of deep narcosis. Her sounds were only sporadic now, but the rattling continued, a rhythmic, metallic click which could have been made without conscious volition. I was just checking the drugs we might need in case of an emergency, when they brought her in: Morris, our pilot, Hugo and Toni, the blanket-covered body slack between them.

Gilka, our first wild chimp patient ever, was out for the count and lay on the table completely relaxed.

'Heavens, what an incredible animal,' someone said, expressing exactly what I felt. The chimp, with her long black-haired limbs and body, leathery skin and grotesque face, looked neither real nor possible. I examined her left hand which she had injured in her struggle to escape; it was so like a human hand that it made me start. When I dressed it with ointment I felt as if I was helping a child, rather than an animal. Someone perhaps from a different planet, but related to me very closely just the same.

'Do you want to give Pentothal?'

Dr Roy had his instruments ready, patiently waiting until we could give the word to begin. We tested Gilka's eyelids and they were still mobile, flicking as I touched them. Her pulse was 120 a minute, her respirations 32; it meant very little since we didn't know the norm, but all being well we would keep a check and see whether, as she recovered, her pulse and breathing rate simulated that of a human.

'We don't need a general anaesthetic; I would rather try it this way, with local.'

Toni, as usual, was painstakingly careful. His dosage rates could be built up, but not down. If Gilka was too heavily narcotized, her recovery time might well be complicated and long. I began to inject the local anaesthetic solution into the very indurated eyebrow area on the right side. At the first needle entry I expected a reaction of some kind, as almost always happened when interfering in this sensitive field. But Gilka, lying peacefully on her back, made no move or sound. Her breathing deepened and slowed a little as the drug effect reached its climax, but the colour of her gums remained pink and healthy.

'Is there any record of chimp anaesthesia,' Hugo asked, 'perhaps one in captivity?'

'Yes,' Toni answered, 'but an animal confined in a Zoo has given up; it eats and sleeps and sometimes breeds, but the zest for life has gone. Consequently it is far easier to anaesthetize such an animal, since the mental state of the patient strongly affects drug absorption. When there is a built-in resistance to interference and deviation from routine, a wild animal, and especially a highly nervous one such as the chimpanzee, fights to the last and in so doing requires completely different handling. Gilka might have done much better with a higher initial dosage level, but

I could not take the chance.'

She was ready for the first incision. Thank heavens the growth was hard and fibrous and did not require complete resection. It was definitely not a case of Yaws; we were almost certain that the abnormal tissue was either of viral or fungoid origin and had, in fact, been so certain from the original history that we had sent suitable medication to be given with the tranquillizer. The small snip of tissue excised would be examined pathologically at the earliest opportunity. After that a more accurate prognosis could be made.

The stitches were neatly in place and everything was as it should be. From time to time one of the group of young ethologists came in to check progress and to tell us of the whereabouts of the rest of the troop. It was just before 11 a.m., almost three hours since Toni had given his first injection. Gilka was ready to be returned to the cage for the recovery period, which might last a long time. She would have to be watched carefully since any reflex reaction in a confined space could cost her a damaged arm or leg. Douglas Roy took blood, I sprayed the wound line with antibiotic-gentian, checked her hand again, and she was ready to go, covered in a blanket in case of shock. So far all had gone well, but there was more involved in this case than just the anaesthetic, recovery and diagnosis. If Gilka's behaviour pattern changed as a consequence of our interference, she might transmit her fear and distrust to the rest of the chimps, wrecking the hard-gained confidence between man and animal. If this happened the chimp study might be in jeopardy, for nothing could be achieved unless the students could follow their subjects day after day. Everyone was aware of the study animals' potential danger if aroused; if they resented human company they might attack and become extremely dangerous.

I decided to take the first shift. Gilka, lying comfortably on her side, was snoring, her long-soled foot twitching in her sleep. Hugo, sitting quietly outside the cage, puffed away at his cigarette, pleased and relieved that, at least, the surgical drama was over. He planned to dismantle the cage as soon as Gilka was free so that all traces of unpleasant association would be removed.

'The most marvellous part of all this,' I found myself saying to Hugo, 'is that within twenty-four hours Gilka will walk out of this cage and be free.'

Hugo nodded and smiled and I sensed that he, too, was thinking of the chimps who would never be free again.

Was I terribly wrong about zoos? Perhaps, after all, they were necessary for those who were not as fortunate as we were and whose only contact with the animal world was in a zoo. Films, television, books of photographs were only substitutes, for there is nothing to compare with the experience of seeing the real, living specimen before your eyes. Modern zoos, like Whipsnade and San Diego, with ample space and water-pools and a corner in which to hide when the peering eyes could no longer be tolerated – would not a zoo like that be less oppressive?

All this, of course, was looking at the problem from our own, selfish human point of view from which we seem not to be able to escape. We wield the power – though this does not necessarily give us the right – to enslave animals, removing them permanently from their own habitat and social groups to a life of captivity.

I looked down at Gilka, then timed her respirations and her pulse. They were slowing down. Toni had brought me a huge beermug full of some punch-like looking lemon tea; Hugo had gone to get breakfast, and I was left alone with my patient, who seemed so childlike, so vulnerable; despite her features she was not devoid of the wild grace that is possessed by those who claim the wilderness as their own.

I thought of a chimp I knew in a zoo, and remembered how he had been jeered at and teased and how his audience had laughed at his antics. He had responded rather like the broken-hearted clown in Leoncavallo's great opera, *Pagliacci*: playing his part with animated gestures, his eyes immeasurably sad.

And what, then, of dignity? We speak grandly of preserving human dignity and rights at all cost. 'We are all born equal under one universal law,' is their cry. 'We must see to it that everyone is treated alike.'

Have animals, then, no rights to defend? To them, life is surely as dear as it is to us. To say that they need less because they think less is rationalizing away our responsibilities. Here, in this beautiful corner of the earth, life was their own – as yet. A few enlightened people attempted to protect the rights not only of the chimps, but of the other fauna. How long would it take before burgeoning humanity felt it must usurp this minute corner of a huge continent?

Gilka stirred and tried to sit up, her eyes tight shut, her tongue licking at her dry lips. Hugo had come back with a bowl of water and slowly I dripped it into the side of her mouth. She swallowed and we helped her up so she could lean against the side. Although she seemed to be recovering, her pulse and respirations slowing down to human pace, she was far from being herself. We doubted that she would be fit to leave the cage before the next day.

'Come on, time to change shift and let's go and swim in the lake.'

Jean, a Cambridge student, had come down with Toni to do her stint. She decided to close the cage door and to sit outside; someone was watching the troop and every time a chimp approached, there would be ample warning. Flo and Malisa, Gilka's special girl-friends, had come to inspect. They sat at the entrance of the palm cover for a time as if considering the situation; was Gilka in need of help – did she like being there – what was she doing, anyway? I glimpsed their faces and their postures through the leaves and wondered. When Toni arrived they regarded him very seriously as if assessing him, then went off to attend to other matters.

Toni and I took the steep walk down to the shore to our sleeping-hut, changed, and dived into the cool crystal-clear water in record time. The sand and smooth pebbles and wave ripples made it seem very much like the sea, except there was no salt. The lake, a mile deep – the second deepest in the world – tasted fresh and sweet. We lay on our backs in the water and looked back at the steep hill and mountain country, the sheer rises of thick lush palm nut and countless other trees. Everywhere one looked there was abundance – of bird and animal life, of insects and of food. Had there not been a vast lake it might have been very humid and hot. As it was, a cooling breeze blew up from the valley most of the day, turning this jungle retreat into a Garden of Eden.

All that night Gilka slept the sleep of the just. She was no longer under the influence of drugs and slept normally until daylight. Toni and I rose early and reached the cage just before dawn. Hugo was there and together we sat in the tropical darkness waiting for the first light of day to challenge the fading stars. If only we could always see the sunrise and inhale the sweet quietude that this hour exudes. Not in a city, or even a small town where milkmen and dogs and cars disturb the peace, but in a place like this, where birds and falling water and whispering leaves are the only sounds that stir the morning. Just as the robin-chat began his ascending warble the first rays of soft pink lit up the summit over the forest: gradually the dark purple tones of the horizon softened as the day dawned, slowly and without hurry, the sun hidden from view behind the ridge of the hill. I moved nearer to the cage to watch Gilka; as soon as she was fully awake we would reassess her situation and decide whether she was ready to be released.

We knew little of the effect of the drugs we had used in relation to memory; would she remember that she had been tricked into confinement and that she had received a sharp needle prick? In spite of anaesthesia, did she, as human patients sometimes claim they do, bring into consciousness a memory of what had befallen since passing out? Would she be aggressive, terrified, or calm when she shook off her deep sleep; would she be ready to be released back into the forest where she might elude the efforts of those trying to follow her? And how would the other chimps receive her, now covered in human scents and even stranger things?

I switched off my torch; there was just enough dawn-light to make out the little chimp, lying on her side, head resting on her folded arm. She was chewing a little and twitching, but her eyes were closed. It would be a pity to wake her, yet this was the best moment for a confrontation, if there was to be one. The other chimps were still curled up in their nests and if Gilka gave the alarm we would still have those moments we needed to open the cage door. Just then she stirred, perhaps because she felt my stare; as she slowly rose up I moved behind the screen and looked through it. She did not make a sound; she walked round the cage, exploring it, but without any sign of fear. She seemed quite strong, but it was hard to tell how strong in such a small space; her eyes seemed bright, her demeanour cheerful. When Hugo approached her to try to feed her bananas, which she refused, she made only a small sound which might have been surprise at seeing him there. Either she was still sedated mentally, or had lost all fear. We decided that she was ready to be released, hopeful that unless she suddenly reverted to her pre-anaesthetic panic behaviour, she would not rush away out of sight the moment she walked out of the cage. And so it transpired. Hugo pulled at the hidden rope behind the screen and the door opened; at once Gilka emerged, and without hesitation ran up the slope but only a little way. As if the effort had been too much she then sat down near a big tree to rest. One of the males suddenly emerged and approached her and for a moment we were terribly afraid that he might harm her. But after a cursory inspection and a little grooming, he returned into the forest. By the time we went off for breakfast and Lori, the American student, had taken up her vigil, a number of chimps had come and gone, none of them apparently finding anything which caused them to give obvious signs of alarm.

We were finishing our coffee and planning the day when we suddenly felt rather than heard the impact of an approaching group of chimps. The gentle, mystical forest echoed as three males, the lords of the jungle, approached, the largest, known as Humphrey, hurling rocks nonchalantly before him as he cascaded down the slope.

'He is top chimp here,' said Hugo 'though some of the others are even larger. He throws rocks just to show his exuberance and his dominance.'

One of the rocks had barely missed one of the students, but she stood her ground. Every moment and movement of chimp behaviour meant something special and would be recorded. Hurling stones was just as worthy of recording as making love.

When Humphrey had finished his exhibition he and the other two males, Faban and Figin, moved to the centre of the clearing. Perhaps they were awaiting bananas, we could not be quite sure, for they did not beg or entreat in any way as so many other animals will under similar circumstances. Then Faban stood up and strode past us with swinging gait, slowly and proudly, his back very erect, fixing us momentarily with his jet-black eyes. Humphrey and Figin did not move; they sat on, surveying their domain and us with majestic graciousness, they the proprietors, we the trespassers.

For a moment I felt envy for the devotees who lived among this population of remarkable creatures. It was often arduous and exhausting to keep up with the agile jungle-dwellers on their own homeground, yet there was no other way of studying them closely. Our experience with Gilka and the consequent disruption of her normal life were also steps towards new knowledge: the pre-catching cage acclimatization period, the narcosis, biopsy and recovery – and, most important, her behaviour once all the effects of sedation had worn off. Gilka would recover; her growth, we hoped, would gradually shrink and perhaps eventually disappear altogether, with suitable medication, once the exact nature of her tumour was known. She and her brother, the only known pair of siblings among the study group, would be watched and followed for years to come, which would in turn yield biological information that could assist in the conservation of other species – from lemurs to lions.

Chapter Seven

LION IN DISTRESS

The project of one of the greatest authorities on lion behaviour, ex-warden George Adamson, was drawing to a close after four years. He had released and most successfully rehabilitated a group of seven lions into the wilderness of Meru National Park: one adult, two sub-adult and four five-month-old cubs.

While he quietly went about the task of nursing the lions back to freedom, the debate concerning the rights and wrongs of his project raged far and wide. Some hard-bitten hunters pronounced that the only good lion was a dead one; the census-takers claimed that there were already far too many lions in Africa without the addition of the Adamson menage. The faint-hearted thought that half-tame lions, already accustomed to man, might pose a threat in sanctuary land which was visited by humans. The over-emotional said it was cruel to expose hand-reared animals to the wilderness where danger lurked behind every tree.

Only a few really considered the lions as *individual creatures*, as hungry for freedom and as oppressed by confinement as man is himself. George, the focal point of their lives while they reorientated themselves from captivity to freedom, watched over the critical acclimatization with infinite care. Within six months they were protecting their own territory, within one year they were making the first attempt at taking their own prey. Some mated with wild lions and bore cubs in the natural seclusion of rocky strongholds, guarding their privacy even from their mate, yet permitting George, their liberator, to share their maternal bliss. They formed a strong pride unit which other indigenous prides, aggressive to the intruders who invaded their terrain, soon came to respect. When one of their number, the patriarch Ugas, was injured, they lost nothing of their strength and nor did he, though only one eye remained to him.

Toni and I became associated with George's work because, as veterinarians, he needed our help, first with Ugas who had injured his eye which we eventually had to remove, later, with others of his wards, who, according to the critics, should have been left to do or die once they re-entered the wild.

There was one important fact, however, which they overlooked: each case we treated, each drug we used, each anaesthetic we administered, taught us something more not only about one single animal, but others of the same species. George's records of his lions' behaviour, of their day-to-day lives as they merged back into the wild, were invaluable because everything he recorded was *new*. This gentle forceful man, whose home was the bush country and whose companions were its inhabitants, was doing more for large cat conservation than any known man in Africa, with the probable exception of George Schaller, the renowned biologist.

We knew that George Adamson would have to leave Meru, his favourite haunt, and his beloved lions very soon and we felt for him: the lions had become his family. Even when he had not seen them for weeks or months they always knew him and acknowledged him as if he had never been away. He would try to return from time to time and watch their progress, yet already other countries were clamouring for his knowledge and experience which were famed throughout the world.

At the eleventh hour Fate had intervened in the form of a buffalo that Boy, second senior male of the pride, had been attempting to kill. He and his sister Girl, originally mascots of the Kenya-stationed Scots Guards regiment, had won fame for their part in the film *Born Free* and had afterwards been the first to join George at Meru. Both had adjusted to their new-found freedom as ducks to water and Boy particularly had excelled himself as a 'bread winner' for the pride, killing adult buffalo with great expertise.

Now, but a shadow of his former golden self, he had been discovered accidentally not far from the new Parks headquarters in a desperate, emaciated state, in which he must have been for some time already. Our first news that something was wrong came across the radio-telephone transmission service from Parks headquarters, this time not directly but relayed by our own Nairobi telephone exchange which seemed to have less difficulty in receiving.

As soon as I heard them say: 'Is that 26177, there is a radio call for Doctor Harthoorn,' I made frantic signals to Toni. A radio call usually transmitted news of an animal in serious trouble somewhere, and when we received such signals lightning decisions had to be taken; that is, *if* the message was intelligible at all. I heard a faint burr, and asked to have it repeated.

'Please, madam, you must say "over" when you finish your sentence,' came the pitying voice of the operator.

I apologized and asked again, this time adding the magic word 'over' at the end. Still I could not quite interpret the sounds which rolled and rocked across my eardrum, but at least I did catch the word 'Adamson' and 'lion trouble'. It was all a little disconnected, so I decided to plunge in.

'Right, I am now going to repeat what I think you are trying to say. One of George's lions is in trouble with a porcupine quill; and we should fly up at once, over.'

'Yes,' the voice drifted across as if from ten leagues under the sea. 'That is correct, it's Boy's leg, over.'

Toni was writing something on the pad for me, which was as indecipherable as the sounds on the telephone. Then he held up his hands making seven and mouthed 'morning'.

'We cannot come till tomorrow, take-off seven o'clock earliest, over.'

'Yes,' came the faint voice again, 'very lame . . .' followed by something inexplicable. It was important to try and get all details possible since we would be going a long way and had to know what to expect.

'Operator,' I pleaded, 'did *you* get that by any chance?'

'Yes,' said the pitying voice again, 'he said: "the lion is very lame and he is glad you can come tomorrow morning".'

'All right then, see you tomorrow, over,' I said and thanking the patient man for his help, I hung up.

Toni was standing next to me with an expression on his face associated with momentary exhaustion following upon the making of split-second decisions.

'Let's finish our lunch and work it out,' he said; 'if it is just a porcupine quill it shouldn't be too difficult, that is, if the quill is whole and hasn't broken off inside.'

I remembered my canine patients who had been foolish enough to tackle a porcupine; once a fragment entered the limb it might require major surgery to extract it.

But what on earth had Boy been doing with a porcupine?

Toni was taking afternoon laboratory classes at the University and left me to check our medical cases and hire a plane to fly us to Meru; if possible, he had added wistfully, piloted by a man who knew the way to the new headquarters airstrip which had only just been completed. This time I would be armed with the new map given to us recently by Peter Jenkins, Warden of Meru. And this time too, come what may, I would have no compunction about presenting it to our pilot for his careful perusal, even if it did offend him.

The last occasion on which I had flown to Meru Park, unfortunately without Toni, I had been assured by the nettled pilot that he *never lost his way* and that I should stop worrying. As a result of his obstinacy in refusing to take bearings we did lose our way most thoroughly so that an hour later, already 15 minutes after scheduled landing time, he admitted that 'I am sorry, my dear, we are a little lost; can you recognize any of the landmarks?'

I could be of no assistance to him at all since I had been born minus a sense of direction; my only contribution was to send fervent prayers to the patron saint of travellers for an early resolution to our problems. When we finally found the balancing rock of the old headquarters, we were 45 minutes late and I was somewhat green about the gills.

We had no need to worry about our pilot this time. Martin Dowling, ex-farmer, had a quiet confidence and skill which carried us exactly to twin-peaked Mugwongo Hill, the site of George's old camp which, but for the fence, had already been demolished in preparation for George's imminent departure. We made a smooth landing on the firm earth runway, and a few minutes later George, stripped to the waist, drove up in his Land-Rover, smiling and relieved to see us, yet at the same time looking deeply worried. Boy, not very far away from headquarters, had indeed been discovered less than 24 hours before we arrived, wounded and completely starved. He had probably tried to eat a porcupine, the only animal he could catch in his battered state, with the result that some of the quills had penetrated his face and limbs. George had been asked to take a look at what the Warden had thought to be a wild lion in trouble and had immediately recognized Boy who had growled ominously at both of them. George, undeterred, had got out of the Land-Rover and approached him, completely ignoring his aggressive behaviour. Devoid of fear, certain that his friend would not harm him, he had walked over and extracted several porcupine quills from the lion's face, very much in the manner of a modern Androcles. When George had done what he could he had requested that a message be sent to us by Parks' radio. Then he went to get meat and water for Boy and stayed with him for the rest of that day and that night. Whatever Peter Jenkins might decide later, the first step was to establish why Boy was unable to move.

Within a few minutes' drive from the airstrip George parked the Land-Rover and went alone to investigate. Boy was exactly where he had left him, lying on his side near a low thorn bush. He suggested that it might be wiser if we remained at a distance to begin with while he injected him with the immobilizing substance we had brought. To us this seemed as dangerous as extracting porcupine quills, but George did not hesitate for a moment. He knew his lions and could predict their behaviour towards him, confident that, even when wounded, Boy would not harm a hair on his head.

Toni gave him the prepared syringe and we waited apprehensively for a grunt or growl or worse that might emanate from the thorny scrub into which George had disappeared. But all remained quiet. After ten minutes George returned and beckoned us on; already Boy had that drunken look which Sernylan induces, his pin-point pupils trying to focus, his magnificent gold- and black-maned head weaving unsteadily as he took comfort from George's voice.

'Boy, old Boy, poor old Boy . . .' George grunted the words softly and the lion responded to him as if he was seeing him from a great distance, moaning and groaning as George soothed and fondled him like a child. Gradually, after a few minutes more, as if unable to continue the unequal battle, Boy's head sank slowly on to his left side, and stayed there. It was sad to see how he had altered; on the last occasion we had seen him six months previously he had been a powerful, vigorous jungle-cat, romping with Suswa, one of the elder cubs round Mugwongo Camp. George had always predicted that Boy would grow into a bigger specimen than Ugas and he had been right; yet looking down at him at that moment I wondered if his injured, shrunken body could ever regain that superb, lithe muscle power and regal grace which had distinguished him from others of his kind.

I had already prepared the general anaesthetic which I now injected into the clipped clean tarsal (leg) vein for complete relaxation, for we had yet to ascertain the exact extent of his injury. Was it really only a porcupine quill in his leg that so handicapped him? The front upper limb looked swollen and displaced; I shuddered to think what else we were about to find.

Martin Dowling, the pilot, had offered to help and we were most grateful, for turning over a lion weighing some 400lb. on to his other side was quite an undertaking. George had meanwhile discovered another quill in the right foreleg; but this had been easy to extract, so easy in fact, that as I levered with large artery forceps, the broken half suddenly went straight from Boy's foot into my own thumb with the impact of my weight behind it.

The right leg was not only swollen but badly bruised with an ugly deep wound between the shoulder and chest. I had brought a small bucket half-filled with disinfectant solution. I now clipped the hair and washed the injured area with surgical soap, then explored the opening with two fingers, following a deep channel which ended in a mobile, crepitating bone fragment. While Toni slowly manipulated the leg from the outside I could clearly feel the 'give' and 'grind' of the fractured limb from the inside. I placed my free hand under the leg and we repeated the movement, trying to pinpoint the exact direction of the break. The least that had happened was a fracture of the upper foreleg bone, the humerus, which had broken across in the upper third. Without X-ray we could not be certain of the extent of the fracture but hoped that it might be a simple break, not that any injury of a wild adult lion could ever be classed as simple.

My heart bled for George who had suddenly gone very pale under his deep brown tan.

'Can anything be done?' he asked.

'The only hope is to fix the break internally, which is a very long operation; a plaster cast is out of the question,' Toni said.

In addition we had discovered a small abdominal hernia which would also have to be repaired. Would Boy in his present state be able to stand the long anaesthetic time? Surgery to repair the fracture could only be done here, in the bush, where there was no chance to establish the exact line and extent of the break by X-ray. Even if we did try to pin the bone, how would Boy react to nursing afterwards? Would he tolerate extensive handling and medication and months of close confinement? There were endless 'ifs' and many hurdles to clear, the greatest at the moment being to obtain the Warden's permission to do anything for Boy at all.

'We'd like to try,' Toni said. 'It will take a long time to heal even if surgery succeeds. Will you be able to give him constant care for several months now that you are about to leave the Park?'

'I'll find a way,' George said, deeply moved at the sudden hope. 'But I would rather shoot him here and now if nothing can be done for the leg. I don't want him to limp in a Zoo for the rest of his life.'

'Let's go and see Peter,' I suggested. 'It all depends on him. Let's see what he says.'

'Don't be too hopeful,' Toni warned. 'In most cases a lion with a broken leg would be destroyed without question. We can hardly blame Peter if he takes that line.'

The Warden was absorbed at his desk as we entered but he put away his work, got up and welcomed us.

'We've come to give you our report on Boy,' I said, 'we won't keep you long.'

Peter was always overwhelmed with work, for his was a marathon task. He not only looked after Meru but also Marsabit Game Reserve, an hour's flight away to the north. He had just come in from a duty trip and was trying to catch up on his administration of the Park. Up there two tribes had been warring and Peter had had the task of protecting his own rangers and checking their posts. When he returned to Meru there was no time to rest and, although he was strong and very capable, emotion and fatigue were there in his face; a slight paleness below the eyes and a tight set of the jaws as he spoke.

'Thanks for coming. What do you think?'

Toni gave him the full picture and summarized our findings.

'It's a clean break, or seems to be, of the upper arm bone, the humerus. It's over to you, but if you will let us, and allow aftercare, we would like to have a go at pinning the leg, of putting intra-medullary steel pins into the bone-marrow cavity. It may not work and in any case the break may be more complicated than we think. Whatever happens, we will have learnt something from trying. And we've got to do it here in Meru. We can't hospitalize Boy.

It's here or not at all.'

'Any other injuries?' asked Peter.

'Yes,' Toni continued, 'an abdominal hernia, but not serious. We could repair that at the same time as the fracture. Trouble is we have to find the right equipment, very strong pins and all the tools for this kind of surgery. Might have to fly them out from UK or US. Then there is the anaesthetic risk, too, but that's always there.'

There was a long, tense silence. Peter was weighing it up, perhaps regretting he had allowed us to come at all. We had told him about the porcupine quills and Boy's wasted condition, especially of his shoulder muscles. This indicated that the injury must have been caused weeks before, perhaps two, or more. It looked like a buffalo-horn thrust, the wound was that kind of shape – and length. It needed the mighty strength of a buffalo toss to split such a bone.

'All right,' Peter said at last, 'how long till you know if it will work?'

Toni and I looked at each other, communicating silently. 'Three days for the primary shock and anaesthetic to wear off. Then about two weeks until we have some idea if the leg *can* ever be sound. But since we have no hope of X-raying it there is always the element of the unknown. Would you give us that much time?'

Peter, looking out of the window as if to gain inspiration, was faced with a very difficult decision. George was due on leave by the end of the year, yet with a convalescing lion this might be difficult. Finally, after some moments of indecision which must have seemed like an eternity to George, he gave us his answer:

'Okay,' he said crisply, 'what else?'

'He needs to be cared for until we can fly up and operate. Feeding, antibiotics, careful watching.'

'That's all right too. But I make this condition: that you and George don't have more contact with him than you need to.'

He looked straight at George who was bridling and suddenly angry. To him contact with a sick lion meant nothing unusual, but to Peter it meant potential danger, to human life. Boy lived near his own home and the new roads to Meru Park were opening up. Boy could become a threat if he was free and could not care for himself.

'If after the time you mention you feel there is no hope of complete recovery, Boy must be shot or removed out of the Park.'

'All right,' said George, 'that's fair. Thank you.'

'Where will you take him now?' asked Peter. 'He's still asleep, isn't he?'

'I'd like to stay with him till he is round from the drugs then drive him to Mugwongo. I can take him in my caged pickup.'

When we left in the early afternoon, having moved Boy to a shadier place, George looked almost happy. He had won a chance for Boy, though at present that was all. When we returned a few days later he related how, that night, as the lion slowly recovered from the drug effects but had trouble sitting upright, George had supported him, a bottle of whisky by his side, his pipe in his mouth. Through those hours, sitting back to back, two lionhearts in the bush, George's world must have seemed a little hazy, a little unreal, even for him; and all the while, leaning on his human friend, the lion's world had slowly returned into focus.

Chapter Eight

THE LONG SLEEP

The flight back to Nairobi was smooth and uneventful and gave us a chance to reflect on our new commitment. It was so very different from all previous cases – even the lion, Ugas – since time was the vital factor whichever way we looked. The Warden of the Park was adamant that George's project must be concluded by December, at the latest, and that the rehabilitated pride should, thereafter, have no contact with humans at all. This was the end of October; even if the operation succeeded and the prognosis was favourable, could Boy recover the use of his leg in two months when it would take a human patient six? If he was still lame at the end of the year and George decided to persevere (which we knew he would) he would then have to move Boy out of the Park for the rest of his convalescence. He would certainly be refused re-entry once he had recovered and George would have to find a new wild area into which to rehabilitate him. Perhaps it was best not to think too far ahead before we had even begun and we certainly could not begin before we had the wherewithal to do so.

'Let's skip lunch and go straight to Invalid House,' Toni suggested as we were about to land. 'If they haven't got what we need no one else in Nairobi will. Then it means importing the pins which will take endless time; but perhaps this is our lucky day.'

I was wondering where, if we drew a blank in Nairobi, we could buy two enormous stainless-steel pins, large-lion strength. Would a zoo veterinarian or a circus doctor in Britain or America be able to help us? Did such a critical lion injury ever receive surgical treatment in normal circumstances? Perhaps, from the economical point of view, it just wasn't worth the time or money and from the cosmetic angle it was also a poor bet since the audiences of the world, film, circus or zoo, were always horrified at the slightest abnormality in a wild beast.

As we drove towards the medical supply house I took a look at myself in the mirror and confirmed what I felt: dusty, hot and tangled, my knees covered in earth stains where I had knelt on the ground cleaning up Boy's wounds. We had been in such a hurry to get back to Nairobi before closing time that we had forgotten to wash and change. I hardly felt equal to plunging into the sterile atmosphere of a shining medical sales room, but I need not have worried. The person in charge appeared not to notice such details as dust and perspiration and when we told him our needs he did not bat an eyelid. He squatted down on to the floor opposite a long glass-fronted row of shelves and began to search.

'Fourteen inches did you say?'

We had carefully measured and compared the sound forelimb with the broken one, taking into account the two-inch shortening. From among the rows of gleaming instruments he now extracted two pins which were hollow and open at one side, like a clover-leaf.

'Excellent,' said Toni, as he examined them with surgical reverence. 'Just what we want.'

He was already planning the mechanics of the carpentry involved, rather like a small boy with a new and exciting addition to his tools. We needed a drill and mallet, chuck, rope and pulley. We might borrow one or other item but knew from experience that bush surgery took severe toll of tools and instruments and that it was far wiser to have our own. There were so many things to think of, so many contingencies, that it made my head spin. We could not afford to leave out one single item, however trivial; for once we were back in the bush, we were on our own and there was no way of gathering reinforcements. We checked and double-checked and talked through the operation plan from A to Z. Of one thing we were absolutely certain: our own four hands, fully occupied *and sterile* during surgery would not be enough. We needed four more, at least, and skilled hands at that, if we were to complete the task in minimum time; certainly not less than three hours. Our friend, the veterinarian, Dr Paul Sayer, who helped us a great deal on another occasion, could not come with us at such short notice. George would be there at the pulley, which would extend the overriding bones, but someone else would have to keep constant watch on respirations and anaesthetic. We very much hoped that this time we would not have to operate at floor level, as always in the past. George had promised to try and rig up something for us under cover to keep off the sun; if he succeeded it would require many strong arms to lift our well-fed patient on and off the table.

Well fed!

As soon as Boy recovered he would be eating 10–15lb. of goat meat once a day at least, and here we were fixing D-Day for Sunday morning, only two days ahead!

Our patient would have to be fasted, not feasted, on the day before surgery. George was administering antibiotics to protect him and hoped he would be able to persuade him to drink large quantities of glucose water to restore his dehydrated tissues. It would be a pity to have to reduce food intake, but on balance the consequences of regurgitation during prolonged anaesthesia could be far more severe.

We composed a radio signal for George and sent it via the Parks' network:

'Do not feed Boy after Saturday morning. Arriving Saturday four o'clock. Operation set Sunday dawn. Give plenty of fluids.'

Even if this message did get jingled up we thought that George, with his wide experience of animal nursing and his vast common sense, would guess at our meaning. He was, in fact, the kind of client every veterinarian dreams of: thoroughly factual, not over-emotional, and a wonderful nurse. If ever a man knew how to ease and comfort his patients and inspire them with courage and confidence, then it was Bwana Simba, the lion man, George Adamson.

There was still one thing to be done before leaving for Meru: the humerus (upper bone of the arm) of a lion had to be found and X-rayed so that we would have something to go by anatomically. We searched high and low in the museum anatomical department and at last unearthed the humerus of a young female lion who had died at Bristol Zoo. Of all the lions shot in Kenya, not one set of bones had ever been scientifically preserved.

But it was better than nothing. We looked at it very carefully with the aid of an anatomy book, comparing the norm with what we knew of the abnormal bone we would be working on. We tried to work out the best method of extending the contracted limb with the aid of a pulley. After our museum humerus was X-rayed we took a good look at the medullary (marrow) cavity and tried to visualize the angle and entry point of the pins. Whichever way we looked at it – and we fully realized that Boy's humerus might be very different – it appeared complicated. We might find something completely unexpected but in the circumstances there was no alternative. Boy could not be moved into the city and hospitalized; if he were, he would almost certainly contract feline enteritis, of which so many wild cats died. Quite apart from this, there *was* nowhere he could have been moved to, for no hospital would accept a fully-grown lion as a surgical case. It would be far better, even at risk, that he stay in his own environment, with George to care for him. If the operation went badly and there was no hope for the leg, George would destroy him there and then. He did not believe in amputation, or in reducing a free-living lion to the indignity of captivity.

As arranged, we flew back on Saturday afternoon, into a lovely clear and sunny sky. I shared the rear seat with the medical cases which could not fit into the luggage compartment, and which, as always, exuded that tipsy aroma of surgical spirit as the container oozed slightly in consequence of the vibration. I loved the scent but it wasn't to everybody's taste; some of our previous fellow passengers had, in fact, turned slightly green at the first hint of it, confessing afterwards that it had conjured up terrible visions of hospital corridors and white-masked surgeons which they found intensely depressing.

'We'll be there in 45 minutes, all being well,' Martin Dowling, once again our pilot, sang out cheerfully as he gained height. He was dressed very differently this time, uniform discarded, replaced by workmanlike khaki slacks, green shirt and soft hat. I had warned him that we were going to accept his offer of assistance and would he please come prepared.

'You need a good stomach to help on this one,' I had told him. 'The steadiest of pilots might well fold up at the crunch of living bones.'

'I used to farm,' he reminded me, 'and stitch up my own cows. So I'll be fine.'

We had our own food and overnight kit with us and, of course, the inevitable sleeping-bags. Peter Jenkins who had taken his family on vacation to the coast had kindly offered us a Parks' vehicle as well as his home for the night.

That was an invitation we would certainly not refuse, I thought as we touched down on to the runway; sleeping in the boma with Boy and George might be a little crowded and anyway it might disturb Boy a great deal more than it would disturb us. Sleeping outside the fence might also be crowded if some of George's leonine friends came to call. All in all, staying at the headquarters would be best if we could make a pre-dawn start and pick up two scouts on the way.

George was waiting for us next to the landing strip, smiling his welcome. We drove with him to Peter's house less than a mile away and made tea. He wouldn't stay long, for Boy was alone and Mugwongo was half an hour's drive away. We showed him the X-ray and Toni explained the technical details of our planned surgical procedure.

'This X-ray is only a guide,' he said, 'we might find Boy very different. But it does give us something to go on, and, whatever happens, by this time tomorrow we'll all know a great deal more.'

George had been very busy in our absence; he had joined three small tables together and put up a protective shade, and had pumped water at the river so that we need not bring any. According to our instructions Boy had been starved all that day; everything, in fact, was ready for the operation. George put down his cup and made ready to leave, then suddenly stopped in his tracks and hesitated. I guessed from his sudden frown what he was thinking: shades of Ugas, moments before we administered the anaesthetic.

'Yes, there is a risk,' I tried to say it gently. 'There is *always* an anaesthetic risk, but we have now learnt so much more from our experience with Ugas and others. This will take longer, much longer; it's a much more serious operation.'

'Yes,' George answered, 'I realize that, but having done it once I know you two can do it again.' If only we had his confidence.

'See you at first light,' Toni said as George got into his Land-Rover, 'and pray for cool cloudy weather and no wind. We don't want dust in those incisions.'

George looked up as if he was making a pact with the heavens and who knows, perhaps he was. Then he smiled at us again and was gone, back to his shaggy patient to spend an anxious night before the morning that was only a few hours away.

Waking in inky tropical darkness at 4.30 in the morning is to rediscover the gentleness of the wild. The heat and harsh sunlight reflected from the African earth and bush, the turbulence of swinging cloud and sudden storm, the brutality of Nature's need for itself, all these are dispelled, at rest. At last, at this hour, the forest and swamp are at peace. The beating hooves fleeing for their lives, the swooping wings, the terror of the spotted cat, the croc waiting over the next ripple, all are allayed for a few short hours. The night prowlers, hyaena and jackal have ceased to send their fearful calls across the star-lit wilderness, haunted by phantom-shapes contrived by the moon and clouds.

The hour before dawn is described by the great philosophers as the time of elixir. Everything is at rest and at its lowest ebb. If one is awake and listening one can hear the breath of life stirring in one's inner self and the murmur of the African earth, like a slow and steady heartbeat, is hardly audible. Though there is silence or near silence, at this hour, the whole world vibrates more than at any other time in our 24-hour span. The sky, lit faintly by the dimming stars, has an ethereal radiance which fades as the sunrise stirs the day to life.

I walked out on to the verandah and stood for a few moments looking into the darkness, listening. A group of waterbuck hovered not far away, as if seeking protection. I breathed deeply, trying to capture the energy of that pre-dawn emanation so that I could draw from it in the day ahead. It would be a long, long day and it would tax every ounce of strength we had. I took another deep gulp of the warm, fresh air, and felt the smallest of wind-tufts against my cheeks. Then I heard a soft step behind me and felt a hand on my shoulder.

'Wish we could just stay for a little longer,' Toni said, giving me the impulse I needed to tear myself away. 'We had better hurry; it's a good half-hour to Mugwongo.'

Martin, our pilot, was up and packed. He had already rolled his stretcher and put his clothes into his kitbag. I put some food on the table – a brief mouthful of breakfast – but just enough to provide some fortification against the day. There would be nothing more until the job in hand was over, for, once arrived, we needed every moment we'd got to beat the heat and the wind currents.

Just after dawn we reached the dip in the road in sight of George's old camp where we had spent so many wonderful days and nights in the past. Now there was only the ten-foot-high wire fence which George had built to keep out his enthusiastic pride – and nothing more, except the temporary shelter set up for us. He was already there, waiting at the last turn before the gate, blending into the dawn-lightened greens of Mugwongo slope so naturally that we did not notice him until the last moment.

'Boy might be disturbed if too many people come near him at once,' George said as we drew level. 'Perhaps I had better go in alone to give him the first injection.'

We were five in all: two Park scouts, Martin, Toni and

I, far too many strangers to descend on a wounded lion. Toni prepared the first immobilizing dose and handed the syringe to George. We could just see the enclosure from where we were and saw him enter and approach Boy's rear-end, his back to us. Again, there was no reaction as the needle went in – no growl, no sudden movement. Then we saw George stoop down over the lion again and gently and lovingly rub the muscle for a few minutes.

After five minutes Toni slowly walked towards the fence to observe Boy's reaction. After nine minutes he came back and asked me to prepare the Intraval Sodium at once so it would be ready when needed. By the time the Land-Rover entered the enclosure Boy was hazy, his pupils pin-pointing, his head top-heavy. Another three minutes and he pitched over on to his side.

I grabbed my medical bag and extracted the prepared anaesthetic, while Toni brought up the vein. Martin, at my side, held the vials down to me so that I could comfortably draw what I needed into the syringe. A little clipping of the hair over the vein – a swab of spirit and I plunged the needle in, drew back the blood for a moment, asked for release of venous tension and began to inject. George, at the head end, soothed Boy, his body relaxing as the fluid entered the circulation, cubic centimetre by cubic centimetre. George's bedroll now served as a make-shift stretcher on to which the limp lion was lifted then moved next to the jointed table which George had already placed under a canvas awning for sun protection.

'What luxury, we have a table *and* a roof,' I said gratefully. I could not imagine how George had done so many things in the short time between Thursday afternoon and today, Sunday morning. But Martin, who was testing the strength of the tables, thought I was joking. How could anyone operate on anything *but a table*!

'Usually it's ground-level surgery.' Toni tried to explain the rigours of bush work.

'But what about infection?' Martin asked, still mystified. 'How can you keep clean if you work on the ground?'

'We do our best, and keep a bucket of strong disinfectant solution by our side to re-disinfect the wound and our hands when necessary. If the wind blows – so help us.'

Toni was draping a double length of plastic sheeting on to the table and cleansed it with surgical soap while I clipped the wound sites. George placed two small tables we had borrowed from the Warden's house within reach and put our cases on to them, knowing from previous experience how essential it was to have everything we needed at hand so that no time would be wasted once surgery commenced. He and Toni were again discussing the mechanics of the operation: where the pulley, which was to extend the injured limb, would be attached, and where the other rope, looped round the armpit to keep the body from slipping, would be secured.

'Let's lift him up.'

Boy had certainly put on some weight in the three days since our last visit. It took all of six pairs of hands to place him, right side up, head at gate end, on the table. His tongue was moving slightly and so were his eyelids. Toni and I prepared the intravenous glucosaline drip, injected more anaesthetic solution into the lateral tail vein, then attached the needle to the drip which hung from one of the awning supports. Then I showed Martin how to inject further doses of anaesthetic into the infusion tube throughout the operation.

'You are most essential to us, as anaesthetist. If you could watch that end, then we can stay scrubbed up and get on with the operation much faster.'

We knew how terribly lucky we were to have an anaesthetist. Like the table, he would ease the burden considerably. The corneal reflex had to be watched for depth of anaesthesia, the mucous membrane for colour to see just how the body was responding to the shock of the anaesthetic and surgical interference. Most fortunately, Bryan Heath, Joy Adamson's assistant, arrived and offered to assist us. And Joy herself, though much handicapped by a recent injury to her right hand, had also come to help. By the time we had finished our preparations, clipped, re-scrubbed and draped our patient, one and a half hours of precious, windless, cool early-morning time had gone. Everything was laid out in readiness: Bryan knew which syringes held the antibiotics which would be injected into the incisions just before closure. I had prepared a great deal of intravenous anaesthetic, alarmed at the amount we had already used before surgery began.

'I'll give a relaxant, don't you think?' I asked Toni. Thank heavens for his experience; this operation would tap all his resources, for Boy would have to remain immobile for hours; and on that immobility depended, in part, the success of our venture.

'Yes, go ahead. Largactil, 200 milligrams.' It would take twenty minutes before the effect of that injection would register.

Toni was scrubbing up while I, already clean, laid out the instruments in sterilized metal trays. Swabs, needles, gut and surface suture were at the ready. The tools for the actual pinning were enclosed in sterile containers, the pins (nails) themselves sealed in sterilized plastic tubing. Drill, chuck, mallet, all in place. I took another look at the table of emergency drugs in case the heart stopped, in case of excess haemorrhage, in case, in case . . . we just didn't know what to expect.

'That drop, coming one a second, must *not* stop. If it does, it means the needle is blocked, and that's serious. Let me know right away. And please, just watch that the skin of the tail doesn't suddenly look swollen over the needle at entry point. If it does, it means the needle is out of the vein.'

'Okay.' He smiled in that puckish, charming way, as if what he had just been asked to do was an everyday, every moment thing, like flying planes. 'Everything fine,' he added, but through his smile I could see his concentration and his understanding of the responsibility.

'All ready? I want to begin.' Toni, gowned and gloved, had located the site of the incision.

I was intent upon Toni's first incision over the fracture site, about midway down the humerus. Some bleeding, but not too much. I swabbed, then continued to enlarge while Toni asked George to manipulate the foot. A little movement gave us the clue as to the exact fracture site. The X-rays had been suspended at the intravenous-drip end of the operation and we kept them in view. I enlarged more, clipped two spurting arteries. Then Toni took over again, moving down between muscles, to the actual bone extremities. I checked Boy's reflexes and called for another cubic centimetre to be injected. I looked up at George, worried and very pale, waiting for the moment when he could assist us. We didn't want him to have too much to do, for it was, after all, his own friend whom we were cutting up that morning. Even the bravest flinch at such a time.

'Move again.' This time, using sterile drapes to prevent contamination, I moved the limb up and down. Toni, deeply immersed in his task, looked grey with worry and disappointment.

'We knew the ends were overriding,' he said wearily, 'but what we forgot was that it all happened weeks ago. There is a false joint and the union is solid. We need a set of tyre-levers to separate this lot!'

He took the long scalpel-handle and began to ease and to separate. Then I took over, tearing my gloved fingers on the fragments. George was moving the foot gently, and I felt the tissues give little by little as I pushed and tore at them. They felt smooth and velvety, perfect in their effort to produce healing, interspersed with the jagged fragments of the original fracture. Although Toni and I took it in turn, the separation took well over half an hour to complete. At last the ends were free and Toni called George over to work the pulley, while Bryan helped to control the armpit rope.

'Pull harder, please,' Toni asked George, as the shortened leg muscles stubbornly resisted the repositioning of the broken ends.

'Right,' George muttered, obviously very worried in case the multiple pulley block exerted too much force and injured the stretching tissues. Slowly and very gently the leg was extended, while Toni guided the fractured ends with his fingers.

I was sterilizing my hands again, checking instruments, getting new swabs out of the drums. New anaesthetic was going in, the drip speeded for a moment, then returned to normal. We had already placed ointment in Boy's eye to protect it from light since the lid was immobile. I asked Bryan to place a patch of gauze over it as well; then, with a sterile cloth, I lifted the upper lip to check the colour. Boy's gums were very pale pink. Two hours had gone by since the first injection and the worst was still to come.

'I've got them apart, the ends of the bone are level. But . . .' Toni faltered and frowned. 'Oh, for an X-ray,' he said longingly.

'What have you found?' I asked.

'There's another fracture, in the lower segment. This will make pinning very difficult. It's a T-fracture.'

I saw that he was flagging. It seemed much more difficult than we had anticipated. I wished I had forced him to eat more breakfast – even if it had been before 5 a.m.!

Toni was now incising over the top of the humerus, where he would drill a hole and insert the round stainless *Staybrite* steel pin. Toni changed bits and hands and finally, with the chuck, forced the steel pin down, his other hand over the fracture, controlling and palpating at the same time to see if the insertion was correct. He was perspiring, the veins in his hands and wrists giantesque.

'Now the second pin.' Bryan snipped the end of the plastic tubing, and Toni, having put on new gloves to replace the ripped ones, took out the Kuntscher, clover-leaf pin which would give added support and drove it in alongside the round one. Having once established a way through, the second effort was much easier.

'Now tap again.'

I took up the mallet and tapped, praying I would manage it. I never had been very good at driving even a nail into a piece of wood.

'Faster!' Toni commanded.

'Slower!'

'Softer!'

'Stop!'

Toni felt the pin at the other end, but thought it had gone too far and tried to retrieve it by pulling. But it would not move, not even a few millimetres. The pins were solidly and firmly in, and now all we could do was to hope that the second fracture had not diverted their path too much.

'Each take one incision,' I suggested. The only thing left to do now was to sew up.

I had already threaded the needles and sewing up was fast work when there were two to do it. Bryan handed me the penicillin-filled syringes and we injected into the wounds just before closing them up. Then gauze pads attached over the stitch line to give added protection to the wound itself. Martin was already cleaning up the hernia site, ready for our second, minor operation, for which Boy had to be turned on to his other side. Again we checked gum colour, eye reflex, drip speed and respiration. All under control, except that Boy was paler, though not grey or blue which were sure signs of danger.

The hernia repair took only half an hour: apart from a few protective injections the operation was complete. We lifted Boy down, unhooked the drip and cleaned the vein site, covering him with a blanket in case of shock. I remembered how cold I had been on a hot day after my appendix operation. We began to clean up the surgical debris, while Joy made some heavenly tea to revive us.

Soon Joy, Bryan and the scouts had to leave. We put the wooden bench under the awning and prepared lunch. It was just after midday, six hours since we had arrived at Mugwongo. As we sat down the sun came out hot and bright for the first time. There hadn't been a puff of wind all morning.

'The gods were kind,' Toni said, sitting down wearily. 'Hope they stay kind for the aftercare.'

Aftercare. Shades of Ugas leaping over the fence on the third day, stitches and all. If Boy did that it would be disastrous for his leg!

I opened our food bag and spread the contents on the gay plastic tablecloth I had brought along. George looked supremely happy and suddenly very young. He seemed to glow with relief that so far, at any rate, all had gone reasonably well.

'There's a lot to be done and we shall have to come up often,' I said. 'Boy may have a lot of pain. After lunch let's go over the medicines he will need.'

George promised to keep us posted – by radio – on developments, that is, if he could leave Boy alone long enough to go to headquarters. He explained that the wild lions at Mugwongo would resent Boy in an area no longer his territory and would make short work of him if they got the slightest chance.

'We'll try and come up in a week or earlier if there is an emergency. But as far as we can see now, Boy should have no immediate problems,' I said, wondering how on earth George would manage all alone.

Having eaten, we were suddenly very tired indeed. I borrowed a mat and Toni a rug and George took the cushions out of his Land-Rover. Thus we lay, between three and four in the afternoon, and rested, the lion next to us, breathing deeply and regularly, showing all the signs of normal recovery. Only Martin wasn't horizontal.

'If I lie down now I won't get up. I'll be fine until I get home.'

He sat in the Land-Rover and read a thriller that kept him awake. It wouldn't do for our pilot to pass out an hour before take-off.

Chapter Nine

FAREWELL TO MUGWONGO

Now the worrying – my own speciality – began. It wasn't only about Boy that we were concerned, but George, alone, without hut or tent, or any amenities whatsoever. He had to depend on others to supply him with food and water, since Boy, once fully recovered from the anaesthetic, might attract the attention of other wild lions and could not be left alone for very long.

Bryan, who had helped so ably during the operation, and who would have been of great assistance to George during this critical stage of Boy's illness, had asked us for a lift on our return journey. He needed medical attention for his knee which he had injured while lifting heavy rocks for Pippa's grave mound. Pippa, the beautiful cheetah Joy Adamson had returned to the wild, mother of many cubs, was dead. She might have received her injury, also a fracture, from an animal she was trying to kill; one could never be certain. We had been out of the country when the accident occurred; when we returned it had been too late to help.

Had Boy been a human patient he would have been confined to hospital for many months. After that, therapy and special care to restore movement of the wasted muscles: swimming, controlled exercise, and a very gradual return to normal life. As it was, confined in an enclosure 35 yards by 20, barred from taking exercise in his own territory, Boy would regain his muscle tone and use of the injured limb much more slowly, especially as he was hand-fed and had no incentive to move. If, at best, the repaired limb showed signs of normal healing progress, where could Boy, homeless and in need of constant nursing, possibly eke out his convalescence?

The Warden, Peter Jenkins, even with George keeping constant watch and providing food, would not agree to

Boy taking exercise outside the boma. A handicapped wild lion could be very dangerous, he said, and turn both man- and cattle-killer. George did not altogether agree. He had, in his years of bush-work and wardenship, seen many maimed, injured lions that lived perfectly normal lives. Since lions hunt in prides he felt that Boy, if wild, would never starve, even if he was not completely sound.

Two days after returning from Meru Park I submitted our report to the Director of Parks, Perez Olindo. A young man, he held his position with great dignity, shouldering enormous responsibilities at a time when the world's eyes were turned towards Kenya's wildlife, both in criticism and admiration. So many controversial issues rested upon his shoulders that they would have shaken a much older, more experienced man. From the cropping of elephants in Parks to the rehabilitation project of the Adamsons, the press picked up each item and sent it round the world.

'Wild animals belong to everyone,' Bernhard Grzimek averred. It did seem that since so many nations of the world helped to preserve and support African wildlife financially, they also had the right to voice their views. When the news of the Adamsons leaving Meru Park leaked out, some journalists responded with an outcry of 'injustice', without having the slightest idea of what it was all about. A wonderful job of work had been completed, and from the outset everyone closely involved had been aware that the project could not last for ever. Meru Park had been only a Game Reserve when George had obtained permission to release the lions; when it had been promoted to a Park the whole situation had changed profoundly. New roads and places to accommodate more tourists were being built, which meant that the minute flow of visitors would increase enormously as facilities improved. The rehabilitated pride, in itself a great attraction, would then be exposed to the foolhardy approaches of people who did not realize that half-wild lions are as dangerous and as unapproachable as wild ones.

Already, towards the last part of his four-year stay in Meru, George was being badgered by complete strangers.

George Adamson, Toni Harthoorn and the author, investigating Boy's injury

The author checking on the patient's condition
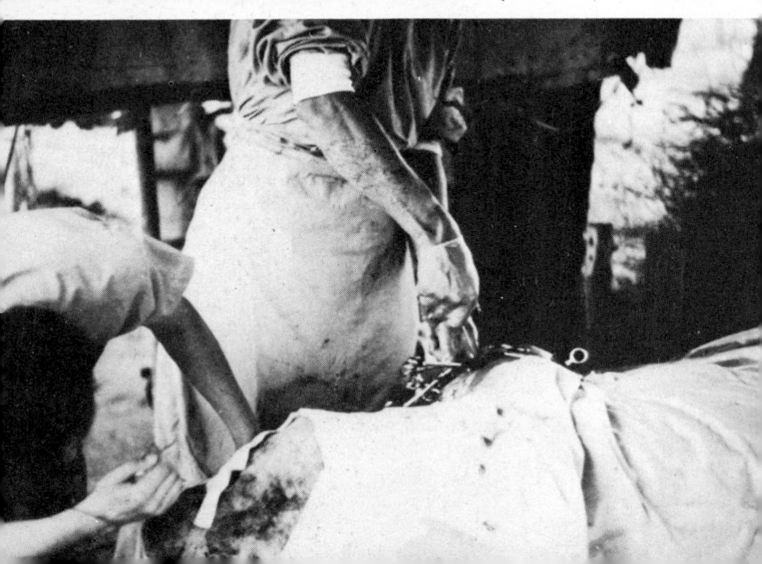

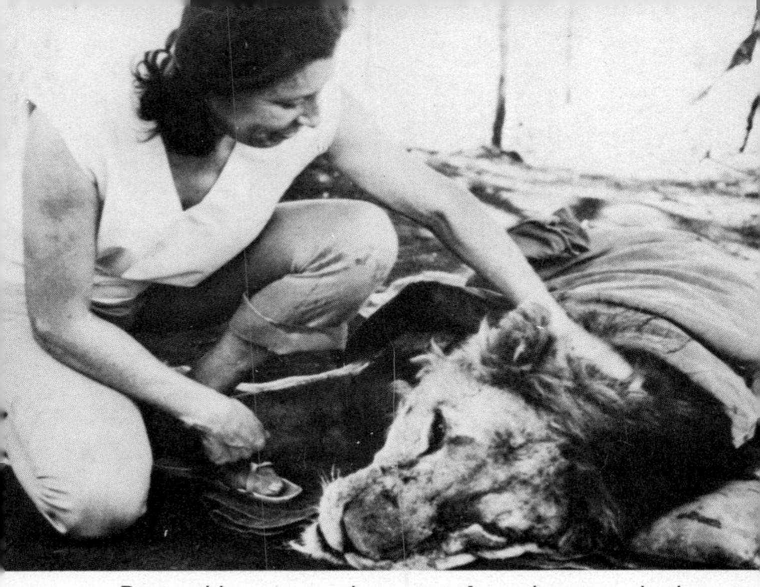

Boy making a normal recovery from the anaesthetic; with the author

Boy seeking consolation from George as the effect of the light anaesthetic wears off

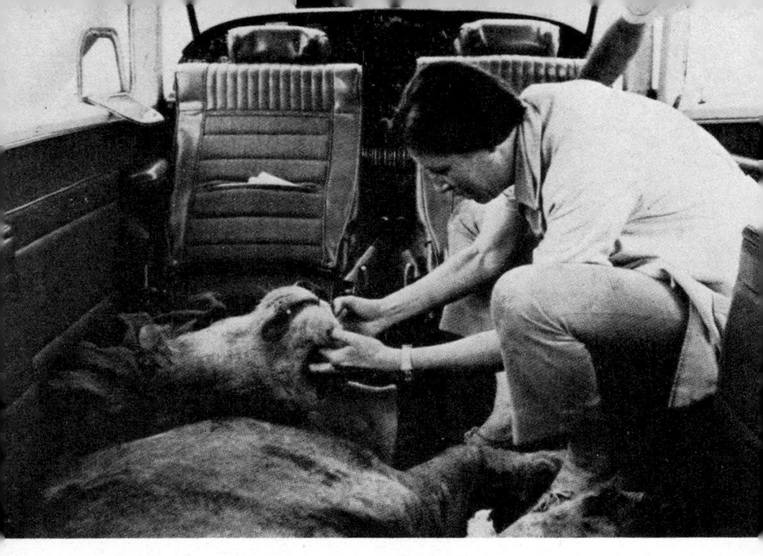

Boy still anaesthetized, examined by the author for ill-effects of the airlift ; with Paul Pearson, the pilot

Boy, still deeply asleep, is moved out of the plane. George Adamson, Paul Pearson, Johnny Baxendale, a kind passer-by and the author

Boy relaxed, but fully conscious, being X-rayed

Boy surveying his new domain

They would arrive and appeal to George's good nature so that they could 'just take one shot with me and that big lion'. They had come a long way for that photograph and could only with difficulty be diverted from their object. When they left, one always wondered if they would try the same technique with a completely wild lion, or with an elephant or buffalo. It seems that when the camera-fever grips the tourist, he will attempt not only the impossible, but the highly dangerous, which at other times he would probably completely condemn.

George, at 63 years of age, was healthy, active and extremely resilient. His store of knowledge was so vast that, whatever happened, it had to be passed on to others before it became irretrievably lost. Meru Park was not the only place where his talents could find expression; and although it would be sad to part from his lions, they now could manage without him.

The Director listened attentively while I described Boy's accident and operation in detail. I had already submitted our report and now showed him some drawings Toni had made which illustrated the actual injury and our repair. He knew how fragile were not only Boy's chances of making a complete recovery, but also the balance of public opinion concerning our interference with Nature.

'This is a scientific project which will help conservation. I will give you and Toni all the support I can, but please, keep me informed.'

When he made his first press release it stated his own view and that of the Ministry of Tourism and Wildlife and the Boards of Parks Trustees. We were delighted that Perez Olindo gave us his full support; but we were dismayed by the flood of publicity which now descended upon us.

'Best to give the press the full story,' the Director suggested when I asked his advice. 'The worst thing that can happen is for the press to be kept in the dark. Tell them the truth so they cannot misreport. You have my full permission.'

Professional etiquette demands that one's name be never mentioned in connection with a case within the country or place of practice. It is a sound rule, though not adhered to in all parts of the world. Basically, it protects fellow practitioners and shields the veterinarian in question from the public eye. There are, of course, exceptional circumstances. But on the whole the press knows that the use of medical professional names is taboo. The Director of Parks, who gave a series of interviews over the months on the Boy case, promised to protect us, though of course we knew that our names would doubtless be used in the overseas press.

As a result of the first official Parks report published three days after surgery, a *Daily Mirror* photographer flew up to Meru Park and somehow found George. On his return he relayed to us the first message we had received about Boy since the day of the operation.

'Boy has fully recovered from the anaesthetic,' he said, to our intense relief. 'The wounds look fine, but he is taking no weight on the leg yet.'

'What a pity we didn't know you were going up,' I said. 'Do you mind if I quizz you for a few moments about our patient?'

'Go ahead,' said the obliging voice.

'Was there much discharge from the wounds?'

'No,' said the pressman, undeterred by the clinical detail I demanded. 'I had a good look, but I don't think there was much.'

'Did Boy seem in pain?' This was most important, not only from the lion's point of view, but from his keeper's. If Boy felt much pain he might well become aggressive and dangerous. In this case we would suggest stepping up the tranquillizing drugs.

'No, George didn't think so and he is giving those pain-killers you left behind.'

'That's good; did you see him move?'

'Yes, we did. He seemed to carry the leg under him, and it touched the ground only a little, now and then, as if he was nursing it.'

'Is he eating well?' I asked, with George in mind.

'One goat a day, quite amazing.'

But the next day, though the newspaper report was good, the photograph reproduced in the local paper was depressing. It showed Boy, as if dead, with an enormous wound. In actual fact, the wound must have looked longer because of the dye-antiseptic we had sprayed on, for the actual incision was no longer than five inches. No doubt he was recovering from a very good meal of goat and was 'out for the count'.

The press reacted at once. Since the photograph, relayed round the world, depicted our patient as, literally, 'on his last legs', some over-emotional under-informed members of the public protested that Boy had been caused unnecessary suffering and should have been destroyed in the first place. Having applauded the effort of saving an individual animal, it suddenly, because of one photograph, became an act of cruelty. The fate of the lion, familiar to thousands who had seen *Born Free*, suddenly became top news and our telephone, already so overloaded that Toni frequently threatened to tear it out by the roots, now rang almost incessantly.

On the eighth day after the operation, when we flew back to see our patient, we discovered the answer that would confound the sceptics. We retaliated by giving the press a photograph which proved, beyond doubt, that Boy was making very good progress indeed.

Bill Travers of *Born Free* fame and George's good friend, had cabled to ask a camera-man to record something of Boy's progress so that he could use it in a conservation appeal film programme. The camera-man approached us and we offered to take him on our first post-operative visit, providing the Director of Parks agreed. In the end it was Toni and I who hitched a ride in Paul Pearson's Aztec six-seater which he could fly well above cloud level on that rainy, squally day. It was the first time ever I had flown so high in a small plane: 11,000 feet, amidst thundercloud and driving rain which did not detract from the beauty of the solid snow face of Mount Kenya rippling in the sun, then again darkening as grey streams of cumulus wrapped themselves round us.

Mike Richmond, also a pilot, who was going to make the sound-track for the film, sat in front with Paul. They were exchanging a few anxious words and glances and I heard them making lighthearted jokes, optimistic that, as soon as we left the first half-hour behind, the clouds would open and let us through, giving perfect vision. The friendly beacon Mugwongo would then loom ahead, and after that, 15 minutes more before, God willing, we could touch down.

When a bad squall hit us so that we were jolted and pitched against our seat-belts, there was Paul, smiling away.

'If he smiles, all must be well,' I thought hopefully, but the cloud was as thick and as dark as ever. Forty-five minutes my watch said, coming into landing time. But where and how?

'There we are,' I heard Mike say cheerfully. He was looking down a minute keyhole in the cloud, which I would have thought was too small for a bird to get through, let alone an Aztec.

The nose of the plane was dipping downwards and I closed my eyes to prevent nausea. The moment I had cut out vision, I felt strangely peaceful. 'Thy will be done, so why worry?' I said to myself silently. 'The dear Lord can't possibly turn us off in the middle of an important job.' I opened my eyes and saw that we were heading down fast and that the cloud ceiling was thinning. Then suddenly we were through and very low down, a startled elephant welcoming us to the good mother earth.

Toni turned back and passed me a Polo mint.

'Looks as if we're nearly there,' he said, smiling with relief.

But Mugwongo, our landmark, wasn't where it should have been. We had expected to find the twin-peaks just below cloud level but had been swept off course as we slid through the 'God's hole' – as one pilot called it. Paul Pearson, not in the least perturbed, now used one of those miraculous devices given to pilots. By making incredible split-second calculations and manœuvres we suddenly found our target on the left, where I myself would never have expected to find it.

Mike took up the map I had given him: the Parks map which had stood us in such good stead before. The new headquarters airstrip was in direct line between Mugwongo and Leopard rock and within minutes we had reached it. Paul veered sharply so that the earth pitched above us for a few moments while he circled, preparing to land. The ground felt wonderful. We had made it, almost on time.

George was in great form when we arrived and so was Boy. He seemed to have grown stronger, looking a little more like his old self. The wounds looked dry, that is, the stitched ones. The wound inflicted by the buffalo was still discharging.

'Please make him walk,' we asked George. He went to his small paraffin fridge, took out some meat and enticed the lion to walk to the other end of the boma. Boy rose immediately he spotted the meat and walked haltingly to the other side. He used his right foreleg very little.

'Now let's see how viable the structures of the limb are,' Toni suggested. 'Please put some meat high up in the tree.'

George cut another piece, smaller this time, and stretching up as far as he could from the added height of a wooden box, he placed the meat neatly on a small branch near the trunk of the tree.

Without hesitation, and to our delight, Boy reared up on to his hindlegs, forelegs against the tree trunk, and with his right paw clawed the meat down from the branch, at the same time fully stretching his shoulder, leg and claws. The shoulder muscles were terribly wasted through long disuse and badly needed to be toned up. This kind of extension movement was just what he needed.

This first test proved, beyond doubt, that the all-important nerves which could easily have been destroyed either by the buffalo or during our manipulations, remained undamaged. This being so, the outlook for Boy's complete recovery was much more hopeful.

The next thing to assess was the state of the fractured bone. For this we had to administer a light anaesthetic, just sufficient to daze him, but allowing his eye reflex to remain strong though his muscles were relaxed. Toni tested and re-tested. The bottom end of the pins could just be felt at the distal lateral end of the humerus, at the right side of the bone.

'If only he doesn't move too much at this stage and they hold for a few more weeks, all should be well. We may have to remove them later on.'

I looked at George and saw a twinkle in his eye and a slow smile spread across his weather-beaten, white-bearded face. 'George,' I said, as the truth dawned on me, 'Boy is keeping fairly still at the moment, isn't he? I mean, he doesn't jump about, does he?'

Everyone, Paul, Mike and Toni, were suddenly all attention. I think it was the way George looked – a sort of boyish mischief written on his features – that drew them.

George puffed on his pipe.

'He's being a marvellous patient,' he said, 'except that he will keep jumping up on to the Land-Rover roof and down again. I don't really know how I can stop him. He started on the third day after the op.'

Toni, still manipulating the limb, went slightly pale. He looked straight across at me and shook his head, hopelessly, resigned.

'We can only hope that the pins are strong enough to hold the weight of his body on impact,' he said quietly. 'And we can step up the sedatives. If he's anything like Ugas, nothing but a knock-out drug will hold him.'

Paul began to laugh and the sound was infectious.

'It's incredible,' he said, 'just a few days after a major operation which would put us flat out for weeks; and here he is jumping about like a rabbit.'

'What else does he do to pass his time?' Mike Richmond asked.

'Just eats his head off,' George said, not deterred in the least, 'and, oh yes, he does nibble my toes at night. The first time he did it I was fast asleep and got a dreadful fright. I yelled out thinking it was a strange lion before I realized it was only Boy, poor chap – he slunk away quite offended.'

Only Boy! What an incredible man George Adamson was.

I got to my feet and took the scissors, forceps, swabs, spirit and mercurochrome from the medical case. The two wounds looked well knitted; the stitches could be removed.

While I was busy snipping and clearing up, Toni examined the hernia site which was filling with scar tissue and had become rather knob-like. It looked a little rough, but did the job. We had been forced to economize with anaesthetic time and had repaired it as quickly as possible. Cosmetic appearances were of no importance to a lion.

While Boy was recovering we ate our lunch, everyone putting what food they had brought on a make-shift table. Overhead the blackheaded weavers were as busy as ever, sometimes tumbling to earth, two and two, fighting and squabbling over nesting sites. Others were repairing last year's nests, making ready for the weaver hens who were due to arrive about two weeks later. Building activities had begun late this year, George said, but the fact that they were now in full swing meant that the controversial rains would soon be breaking, after all.

'The weavers are infallible,' he added, 'and thank heavens for that. The Park is parched for lack of water.'

The weather stations had more or less written off the rains and the farmers were desperate. The time of the season had come and gone, yet who were we to doubt the weavers' message?

Apart from short visits, George had vacated the camp some weeks before Boy's accident and must surely have been conspicuous by his absence, not only to the lions who came to call at frequent intervals, but to the birds. I remembered the steel helmet, always full of water, surrounded by grain which he put out especially for the weavers who shared his camp; and the lizards and mongoose, and so many others who came to partake of his hospitality. Mugwongo Camp had been so much more than an experimental station; it had been a place where all were made welcome and enjoyed the wonderful spirit of quiet camaraderie which George exuded.

'Perhaps the weavers came late because you were not here in September; perhaps they felt their guardian spirit had departed.'

George smiled sadly and thought about it for a moment. His camp having only recently been razed to the ground, this might have been transmitted to the weavers by the far-reaching bush telegraph.

'No, I don't think it matters to them whether I am here or not,' he answered finally. 'Their tree is intact and that's the main thing.'

The end of December was D-Day for George; it was only too clear to him and us that Boy could not possibly have recovered by that time, although we now predicted 90 per cent recovery for the leg. An alternative convalescent station would have to be found, not too near the madding crowd, yet where he could have treatment and care until he was ready – for what?

There had already been offers: the Nairobi Orphanage, but we had rejected this for reason of the infection he might contract there. Many zoos and private individuals had offered him sanctuary and even the Scots Guards had rallied, alarmed by adverse publicity which advocated Boy's destruction. Offering to 'come and get him and take him home', they rose strongly in the defence of their ex-protégé. When I heard this, I had visions of a Guards Battalion storming Meru National Park and leading out their beloved lion under protection of gunfire. I wondered very much if they were aware that their one-time cub had reached immense proportions, measuring over ten feet from nose-tip to tail-tip and weighing well over four hundred pounds, even in his present debilitated state.

George fully appreciated everyone's kindness and thanked them; but he felt confident that the National Parks authorities would not be swayed by public opinion and reverse their decision. Meanwhile he wanted only to be permitted to nurse Boy through the difficult stages of convalescence, accorded the privacy that any human patient would receive at such a critical time. Although he hoped to be allowed to remain in Meru as long as possible, it was wise to consider alternatives while there was still time. His choice fell on the Adamsons' holding on the edge of Lake Naivasha, not too far from Nairobi, and yet not too near. Johnny Baxendale, his former assistant, once more came to the rescue; he immediately dropped his work and supervised the building of a large enclosure at the edge of the lakeside woodland. It wasn't as wild as Meru, but it was a beautiful site just the same.

Chapter Ten

THE AIRLIFT

The blackheaded weavers had, as always, been correct in their weather prognostication though they themselves might have been surprised by the deluge of rain which suddenly burst upon the drought-stricken world of north-eastern Kenya. George, reluctant to come to a final decision concerning his departure from Mugwongo, now had his mind made up for him by force of circumstances. The boma at the edge of Lake Naivasha had become his life-line.

Paul Pearson, undaunted by weather or anything else for that matter, suggested that we fly our patient out, about the same time we had first recognized the fact that the healing of Boy's leg would take many more months.

'You really mean it?' I asked him, trying to assess the seriousness of his offer. Toni and I had been turning over the thought in our minds for some time, but did not think that anyone would risk flying an enormous lion in a small plane.

'Of course I mean it, no trouble at all,' Paul replied. 'I'll just take out some seats and strip the door off the baggage compartment. If you and Toni can keep him still, I'll fly him out.'

We had decided to make a preparatory journey to Meru to take the more urgently needed medicines for Boy and to discuss the idea of aerial evacuation with George. On balance this would be very much less arduous for the fractured leg than eight hours or more of driving in a bouncing Land-Rover over a rough sandy road.

Since the rain had not reached Nairobi yet we were amazed to receive a weather signal just before take-off. The message read that the Meru Park headquarters airstrip was flooded and we should not come that day. Though Paul was itching to take the chance, he could not fly in the face of this advice. We would have to wait until the weather cleared. Once the rains set in heavily, George and Boy might become landlocked for months, which could be disastrous for both of them.

For one whole week we were at the ready, our medical cases packed. At last, after days of anxious waiting and no possible communication, two friends of George's attempted the flight in spite of very poor visibility. They chose a four-seater Piper Cub which would be light enough to take off from the wet landing strip at Meru Park headquarters. In the worst event they would drop the medicines and a letter containing our air-lift plan over George's camp.

On the evening of that Sunday Johnny Baxendale, who had spent many months before working with George, telephoned. They had flown 45 minutes of the way when the cloud had suddenly become very dense and rain squalls had challenged the plane. They were thinking of the tragedy that had occurred on the previous day in Tsavo Park West when a young pilot had crashed into the hillside because of poor visibility under similar conditions. The Piper Cub bound for Meru turned back, in the circumstances much the wisest thing to do.

We now had a problem. The continued rain could endanger not only Boy's health, but also George's, since by now the enclosure must have been flooded with water. An animal just out of surgery, as Boy was, was very vulnerable to any added stress and infection. Even if all went reasonably well, how were we to effect an airlift if George had no warning of our coming? How were we to know that he would, in the first place, agree to our scheme of taking Boy out by plane?

Tuesday, November 25th, Paul Pearson, speaking on early radio call to the Warden of Meru Park, was told that the first night without rain had just passed.

'Try and come today, in case it closes up again later in the week,' was the gist of the weather-distorted message. Paul was to try and contact us, see if everything could be organized at such short notice, and then go back on radio-channel at 11 a.m.

I had made all my preparations and was eager to get going. For the return there had to be someone to meet us at the Naivasha airstrip, and there had to be sufficient pairs of hands to lift the lion out of the plane and into the waiting vehicle. There had to be a vehicle that George could use at Naivasha, which meant extra transport. The camera-man employed by Bill Travers had to be alerted, for Bill was hoping to show a film of Boy's continued progress to confound the 'anti-lion rescue' clan, aiming to arouse interest in George's project, already beautifully portrayed in the film *The Lions are Free*. There had to be food, clothes, lamps, a tent for George, meat for Boy and some bedding. Mike Richmond and Johnny Baxendale offered to arrange everything in time and meet us with a spare vehicle at the Naivasha strip by 3 p.m.

At 11 a.m. Paul Pearson again tuned into the radio channel. The Warden wasn't there, but the assistant informed him that since transmitting earlier in the morning there had been a new rainfall; he was doubtful whether we should come. This time Paul was not to be deterred. Since the Warden was elsewhere, Paul assumed that the assistant did not really understand the circumstances or the fact that Paul was one of the most experienced bad-weather flying pilots in East Africa.

'Okay, I'm ready to leave at midday,' he said on the phone. 'Are you?'

The days of anxiety and waiting had brought me to such a pitch that I thought of nothing else but getting to Meru Park. I suspect that Paul did not realize that I would be alone on the job, Toni being away an an IUCN* meeting in India, and that he was committing the safety of his passengers to a female vet. He had left one extra seat in the plane, so that there were four in all. The other two had been taken out while the rear storage compartment had been opened to form one continuous space with the cabin. We were flying Joy Adamson's assistant, Bryan Heath, up with us, but he would be staying there. On the

* International Union for the Conservation of Nature and Natural Resources.

way back there would, we fervently hoped, be George on the floor with Boy, myself and the pilot. One seat would then be folded and put in with the luggage.

'I think there is one seat too many,' I said on arrival at the airport. 'Toni is away and I cannot reach the vet I had hoped might be able to come with us.'

Paul's smiling, welcoming face grew serious for just a moment and I had the impression that he flinched just a little, but that was all. He simply went over to the plane, climbed in and took out another beautifully upholstered red seat. There was an enormous tarpaulin to protect the floor, everything cleared for added space and easier manœuvrability. He had certainly pitched in with enormous enthusiasm, and I felt sure that he would get us there; I also knew that on that day of heavy rain and thick cloud I would not have undertaken to fly an adult injured lion with anyone else but Paul.

This time our flight up wasn't bad at all for we could actually see the ground, a very comforting feeling.

'You're flying very low today,' I said to Paul.

He smiled, explaining that we were flying at 11,000 feet, the same altitude as the previous week, but the cloud, this time, was above, instead of below.

Suddenly, within moments we had lost the link with *terra firma*, and before us, like a giant ghost-castle, towered Mugwongo, magnified to giantesque proportions by the mist, distorting the familiar twin-peak beacon. As we approached driving whirls of rain hit our windows with a frightening hiss, but Paul was delighted, for once at Mugwongo we were so nearly on target. He circled twice round the peaks, very low, looking for George and his tent and Land-Rover which we could not see. We assumed, rather optimistically, that George had already received news of our intended plan of action and was driving to the airstrip with Boy in the back, ready to be tranquillized for the flight out.

In a few moments we were over the double runway of the Park's headquarters which was not only wet, but covered with pools of water which glistened everywhere.

Paul groaned; Bryan and I looked at each other in despair. Had we really come so far, only to turn back? We circled widely at first and saw two Land-Rovers and Peter Jenkins's Piper Cub tucked into the hangar. As we circled again, this time more tightly (when for moments I could not separate heaven from earth), we saw one of the Land-Rovers start up and drive on to the runway itself.

'I wish Peter would get on the radio,' exploded Paul, 'and give us the runway condition: I am going in low,' he added, making exactly as if to land, but just skimming over the thin strip. Then we recognized Joy in her Land-Rover, making some sort of sign, pointing to the Piper Cub.

'Maybe he can't reach us,' said Paul, raising up and turning back; 'write a note, will you, Sue, tell him our frequency is 18876.'

'Right,' I said, beginning to scribble, 'but how do we float the note down?'

'Anything, anything at all,' said Paul cheerfully, banking to make another low run.

I opened my safari handbag, a sort of hold-all that held all. I saw my spare pair of threadbare tackies and some socks. I fished out one shoe, pushed the note into the toe, tied the sock tightly round it all and passed it to Paul.

'Just the thing,' he said, and swooped very low.

'No one will see it,' I said, wondering how Paul could aim at the strip when we were coming in at a frightening speed.

Just then he opened the little peep-hole window on his side and threw out the shoe, aiming backwards at the very moment when we passed the Land-Rover. A white-overalled mechanic was ready for us and rushed out to gather the missile. Within minutes Peter's clear voice came over the air.

'It's rather wet,' he said, 'though one runway section isn't too bad at all. It's taking off again that might be difficult, if there's more rain. Over.'

'Roger. I'd like to try if you think we can make it. Over.'

'Just exactly what is your plan? George is still at Mugwongo. Over.'

We had feared this when his Land-Rover had not been visible near the airstrip.

'Can we get there by road and get him and Boy in time to take off and reach Naivasha?'

'Yes, you can. I take it you have Sue with you?'

Peter knew Toni was away and so his question was extremely pertinent. He might just for a moment have wondered how Paul actually contemplated flying one enormous lion out in a six-seater plane.

'Roger. Yes, I have Sue here, am coming in to land.'

We were delighted, though we knew full well that if a rainstorm, such as we had just passed at Mugwongo, hit the airstrip that afternoon, the chances were we would not be able to take off that day.

Paul landed superbly – he always did. It was very wet and we were glad to be able to park on the concrete slab in front of the hangar. We made a quick plan and divided into two groups. Joy and I drove straight to Mugwongo so that I could anaesthetize Boy, while Peter, Bryan and Paul went to pick up some scouts to help lift the patient into the Land-Rover. In this way we would save time and give George a chance to think over the idea of the airlift.

The road to Mugwongo, usually a matter of 30 minutes' drive, was as muddy as the landing strip. It took us over an hour to reach George; as Joy and I drove along skidding, engine protesting, weaving from one bank of mud to another, it occurred to me that George might actually prefer to risk staying on in Meru rather than make an air journey into the unknown. He would have no time to pack or consider. It was either yes or no. If yes, then it was now, at once, so that we could reach Naivasha, 45 minutes' flying time away, in time for Paul to take off for Nairobi in daylight. If no, then at least Boy would have new supplies of medicine and I would have a chance to check progress. The next few moments would be decisive.

'I really didn't think you would get down today,' he

exclaimed with surprise. 'I did get the message you might come, but I thought that after that storm you would have to turn back.'

I felt a sudden impulse to seize both George's hands, and beg him to take the risk with us, for Boy's sake, for this would almost certainly be our last chance in weeks to fly up. It was just 23 days since the operation on Boy and he was beginning to take more exercise in his restricted enclosure. The continuous rain and mud would hamper his recovery enormously, apart from the danger of illness he – and George – might contract, under these very adverse conditions. What was needed now was a dry, larger boma and an X-ray machine to establish the extent of healing and position of the realigned bones. Naivasha would provide everything that was needed.

'George, would you be willing to fly Boy out now, this afternoon, to Naivasha? He will have a much better chance there and we can give him more and better attention. I've never flown with a lion, but I am sure we can manage it between us. There isn't time to pack, but everything is arranged for you at the other end.'

George stroked his beard for a moment with that special humble gesture of his. But his hesitation was momentary.

'It's fine with me,' said George, his mind made up. 'I'll do whatever is best for Boy.'

Just then the lion poked his enormous head out of the communal tent, looking at us with those questioning yellow eyes of his. If, at that moment, he had broken into speech I don't think I would have been so very surprised.

I felt intense relief and enormous admiration for George; he was so full of courage and faith, so completely devoid of histrionics. He did what was best, without fuss. If he felt shattered at that moment, as he surely must have done, that his leave-taking of Meru and the lions had come so suddenly, then he did not show it. One of the 'Bisletti' females was mating with a wild lion two hundred yards away. As he was explaining the cause of the uproar Girl arrived out of the bush grunting affectionately at George. He fetched a small piece of meat and threw it to her, after which he went inside the enclosure with the syringe I had prepared. Boy received his first injection at 2.40 p.m.

'Mind she doesn't break your bottles,' he warned me as he closed the gate. I was busy on the bonnet of the Land-Rover outside the fence, making up the longer-lasting anaesthetic solution that was to follow the first drug. Girl, having gulped down the meat, came up to me and gave me an affectionate nudge. Then she leaped up on to the bonnet, neatly avoiding my paraphernalia in true feline fashion, and took possession of the roof from which she observed our activities with a kind of tolerant interest. Had she come to say goodbye to George, or perhaps to her brother Boy, from whom she had never been parted since they were cubs?

I went into the boma and tested the effect of the first phencyclidine injection. Boy was almost relaxed, sufficiently for me to clean up the vein-site in the front legs. I looked up at George: without any obvious haste at all he was collecting a few essentials to take with him: his toothbrush, some whisky and tobacco and very little else. Then he went to the paraffin fridge and turned it off, saying a few words to Twitenguru the faithful tracker who had helped the Adamsons for many years. I could not follow the conversation but guessed that he might be taking his leave of him, too. The Land-Rover arrived just after I had given the anaesthetic injection and the relaxant, which I hoped would tide Boy over until we reached the plane. I could not put him out too deeply, preferring to re-inject every hour or so. On the other hand, he had to be completely immobile to be moved at all, and to fly safely. I thought of Toni and his expertise in this field. It would be terrible if I let him down!

'Promise you won't do this fool thing alone,' he had said just before he left for India. 'Take Paul Sayer with you, for heaven's sake.'

I had promised but at the last moment had not been able to locate Paul. There had been no time to waste, promise or no promise. Even as I worked out the dosage rate for Boy, I could sense Toni worrying. I would send him a telegram as soon as we landed safely.

Within ten minutes, Boy was ready to be loaded into the Land-Rover. While Paul, Peter Jenkins and the scouts wrapped Boy in the blanket I had brought with me, George picked up his small suitcase, took one long look at what remained of his old camp, walked out to the gate to fondle Girl for the last time and got into his own Land-Rover, at the back of which the lion had been placed. I checked Boy once more: his breathing, the colour of his gums, made sure he was lying comfortably and normally without compressing his windpipe. He had, unfortunately, been fed that morning but that could not be helped; George had fed him one whole goat. It meant he would need watching very closely and, above all, his head must not be allowed to lie below body level.

George started the engine as I got in beside him with my medical case. I wore a many-pocketed bush-jacket and while we drove put into it all the emergency drugs which we might need after take-off; I added swabs soaked in spirit, scissors, tape for holding needles in place, needles and syringes. If I were to need any of these in the air, there would hardly be time to search for them in the depths of my overfilled medical bag.

On the way to the airstrip we passed a station-wagon going the other way, amazing, considering the road was almost impassable for anything other than a four-wheel-drive vehicle. We noticed that it stopped as soon as it saw us, turned round and followed, adding to the procession all the way to the airstrip.

'Good heavens,' George asked, 'how did you get up here?'

We should have guessed that only determined pressmen would have made that journey in such weather. They had come, they said, to see George and Boy. Seeing Boy being loaded into the plane was nothing short of a scoop and they made the most of it.

We had hoped so much that we could move Boy quietly and unobtrusively this time, without arousing another 'volley of world opinion'.

'It's uncanny how they manage to be everywhere at the right time,' George commented dryly, 'they always seem to be a few steps ahead. But I don't envy them their long wet drive back!'

We took off, finally, at 4.30 p.m. The loading into the plane had been a slow and difficult process, even with so many people to help. Only when Boy was stretched out on the floor of the cabin did we realize how enormous he really was, and how difficult it would be to manœuvre him if an emergency arose inside the light fragile plane.

George sat on the floor at the lion's tail end, seat-belts fastened for take-off according to the rules, and I sat in front with Paul, my hand on Boy's head.

We had been most fortunate with the weather. Luckily it had not rained one drop since our arrival and the runway had dried out quite considerably. Our take-off was smooth and uneventful and our climb up and out of the Park without any turbulence whatsoever.

At first all went well with Boy; while we ate some late lunch sandwiches, we passed the lower slopes of Mount Kenya and saw the glaciers' white-lit pinnacles just above us. I had re-injected him just after getting off the ground; his breathing was deep and regular, his colour good. George, whom I was also watching carefully, seemed relaxed and fairly happy; at least he did not ever suffer from the discomforts of airsickness. Although I had air-travelled well in the past, I knew that the flight could well become very rough and difficult at any moment and that I had better be in *absolute* control of myself at all times. I had discovered that Stemetil, an anti-emetic drug, did the trick when taken in infinitesimal quantities. No sleepiness and a four-hour security from nausea. Above all, the psychological comfort derived from that knowledge was tremendous.

At 5 p.m., 30 minutes after take-off, Paul was climbing over the Aberdare Mountains into thick cloud and dense turbulence. Suddenly I heard George call my name and as I turned I felt Boy thrash and turn his head in one violent, convulsive movement. I slid to the floor and quickly examined him: his tongue was very blue, his heart over-exerted. Paul looked over his shoulder to tell me we were flying at 12,000 feet. No wonder the lion was suffering from air hunger!

It was too high and in his deeply drugged state he could not stand the additional stress of lack of oxygen. *Oxygen!* Why had I not thought of bringing an oxygen cylinder? It was so obvious that a stressed body cannot cope with sudden changes in altitude. I injected a stimulant and treated for shock.

'Could you please fly lower?' I asked Paul.

'Not yet,' he answered brusquely, busy with the job in hand; 'we're over the Aberdares heading for the lake. If we fly lower now we'll hit something in this thick cloud.'

Boy was holding his own, but only just. His breathing was fast and shallow as he struggled for air. I wondered if I had sedated him too heavily, but knew that I could not have taken a chance and given him less. I looked at his distended abdomen; his circulation must have been stressed to the utmost at that moment since so much of it was already occupied in the digestion of food. I pumped his rib cage slowly but firmly to aid respiration, hoping that the injections were taking effect. *If* Paul did not reduce height soon . . .

'Going down to 9,000.'

Within moments Boy relaxed, his lolling tongue changing from blue to pink, his body untensing. There was no doubt about it, just three thousand feet gave him what he needed. I looked at George, who had turned very pale. His eyes were brimming over with relief.

'He'll be fine now, we're nearly there.' I got up into my seat and put on the seat-belt. George moved up on to the step of the luggage compartment and put on his.

'I can see the lake, we'll be down in a moment,' Paul said. 'Prepare to land.'

Never to me has Lake Naivasha ever looked so beautiful, so solid, so comforting. Paul had spotted the waiting vehicles and made a swift landing on the firm grass strip.

We were two and a quarter hours late and very lucky

that there was anyone waiting for us at all. Although Paul was anxious to return to Nairobi before dark, I felt it best to wait for at least fifteen minutes before commencing to unload our patient. Although three men met us on the airstrip, two of them were totally occupied with cameras, which meant there were only four of us to do the job, as opposed to eight pairs of hands which had been available at Meru. We thought it best not to slide him out backwards through the luggage compartment, the way he had been put in, but move him out forwards, using the reinforced wing section as a slide. It took a long time but we were successful; Boy was still completely relaxed though his eye reflex was returning. By the time we reached the new boma eight miles away, he was just beginning to respond to George's voice. We parked opposite the open gate of the enclosure which had been built at the edge of a forest on the lakeside. It looked wonderfully dry after Meru, but the air temperature was much lower. Naivasha lay at an altitude of 6,000 feet and over 3,000 feet higher than Mugwongo; George and Boy would have to guard carefully against catching a chill.

I decided to wait on and see how Boy would recover. If very slowly and there was any unforeseen difficulty, I would stay the night. Mike Richmond had brought plenty of blankets, sweaters, a lamp, food and coffee.

The main house boasted a telephone which might with luck be in working order and was only a few hundred yards away. Otherwise, the nearest neighbours were half a mile down the road, both up and down the lake.

When, at nine o'clock that night, Boy lifted his head in answer to George's voice, I decided it was safe to leave. Hopefully, on awakening, Boy would walk out of the vehicle straight into the boma so that George would need no further assistance. He would feel very groggy for days and take some time to adjust to his new surroundings. But then so would George. Perhaps Boy's need of George would help to soften the blow of separation from the other lions and from Meru; he would be fully occupied with caring for him, with providing sufficient food, with making plans for the future. But would he also be dreaming of the past and wondering, worrying about his lion pride behind the mountain?

I asked him how he thought the lions looked upon him and whether they would miss him now that he had left.

'I think they look upon me as a sort of elderly lion, a friend,' he said. 'At first they needed me, but gradually, as they became more independent, their need grew less. But their behaviour towards me has always remained the same; once they had given their trust nothing changed. When I go back to visit them, I know it will be just as before.'

Chapter Eleven

HOME IS NOT A FENCE

All the way to Nairobi from Naivasha I wondered whether, after all, I should not have remained behind. George was capable – he had proved it to us on countless occasions – yet to be left alone at the end of such an emotionally and physically exhausting day, would this not be too much, even for him?

'He'll be fine, stop worrying,' John Reader, the *Time and Life* photographer who gave me a lift back, had reassured me. 'There's nothing that you can do out there that he can't do himself.'

When I finally got home and into bed, I was weary to the point of tears, but could not sleep. That dreadful moment at high altitude kept coming back, when I myself had felt the stress of the thinning air as we bumped and tossed in the thick cloud. And George's face, more worried than I had ever seen him, yet resigned. I felt myself rocking in that flimsy little plane all over again, flimsy compared to the vast elements that challenged it. It was hard to believe that the nightmare was over and that Boy, thank the dear Lord, had reached the next stage of his journey to recovery.

Toni! I had forgotten to send him a cable. It seemed ridiculous to do so at such a late hour, and yet somewhere on the other side of the Indian Ocean, perhaps in the middle of a conference, he would be wondering and waiting to hear. I debated the best way to send the message; should I say 'mission completed', or 'had to break my promise, all well', or 'come home soon'? I felt a little groggy, my thoughts hardly under control. In the end I decided on the shortest possible sentence: 'Boy safe at Naivasha'. Toni would make his deductions from that and he would curse a little, but be glad it was over. And when he returned he would grumble that if that was the sort of thing I did while he was away, then I just could not be left alone.

'Can't let you out of my sight!' I could just hear him say it, the last thing I remembered before I heard the telephone jangle me out of deepest sleep.

'Good morning, Sue,' came George's calm, deep voice. George hated the telephone; not only that, he feared it. All the more marvellous that he kept his promise to give me a progress report at the very earliest. Boy was doing fine, he said, except that he had more or less completely wrecked the back seat of the Kombi. He had chewed at it with gusto as he was coming round and George could do nothing to prevent it. It was Mike Richmond's car, a loan that was very much appreciated. George felt a little embarrassed to return it looking as if it had, indeed, been chewed up by a lion.

'He has a stomach-ache,' George continued. 'Early this morning he actually growled at me, something he has never done before except when he was mating and I got too near. His stomach seems to go into spasm every now and then.'

I remembered the goat he had eaten a few hours before we left. The anaesthetic had disturbed the digestive process and so, no doubt, had the flight. On top of all this he might well have swallowed some portions of car seat to add to his distress!

'He'll be fine, but all that meat will take a time to pass through. Is he moving about much?'

George sounded very cheerful, considering the sleepless night he must have had. He described how, at first light, Boy had got up, sore tummy and all, and had walked from the Kombi into the boma through the open gate. He was still unsteady, but very conscious, and he had settled into his new home with absolute calm. When George had put up the tent which Mike had brought, he had settled down like a seasoned camper, lying next to George's stretcher to sleep off the remnants of the anaesthetic.

Toni returned earlier than I had expected; apart from muttered complaints about women who kick over the traces and run amok in the absence of their husbands, he accepted my part in the air rescue very philosophically. We made regular visits to see Boy, taking Paul Sayer, the veterinarian from the Veterinary School clinic, as often as possible. We were due to be away for three weeks over Christmas with the children and hoped that Paul would look after our patient during that time since someone had to be at hand in case of difficulty. The original buffalo-horn wound, a long track through the chest muscles, was still discharging and painful and we feared that a piece of bone might well be lodged there and was causing irritation. Each day George cleaned and treated it with antibiotic cream and each day he palpated the site of the former fracture which had now formed a large callus. Boy limped badly, especially after rest, carrying the limb slightly outwards. The end of the pins could be felt quite distinctly, and seemed to be causing some discomfort judging from Boy's preoccupation with that part of his leg. By the time Toni and I were almost due to leave on holiday, weeks had passed since the operation. The time had come to take an X-ray and to find out exactly how the pins and reset bones were holding up under the strain of increasing movement.

Once again the ever-resourceful Mike Richmond came to the rescue. He found a mobile generator-unit powerful enough for the X-ray which Paul Sayer borrowed from the veterinary clinic. The unit could be dismantled and more or less adjusted to any height and position and connected to the generator brought to the boma fence in a Land-Rover. There would be new sounds and strange smells during the procedure and we wondered very much how Boy was going to react.

At first he seemed to be disturbed by the whine of the dynamo, but gradually he settled down next to George as the tranquillizer, fed to him in meat, took effect. He lay half in, half out of the tent, very aware but tolerant of our activities. The X-ray machine had to be suspended from a tent pole, the X-ray plates had to be pushed under his leg, which hurt the sensitive areas. He growled a little but George consoled him and after a while he even allowed himself to be guided into rolling on to the far side so that the other aspect of the injured limb could also be X-rayed. We had decided to take it very slowly, step by step, so that Boy would not feel harassed. It was barely nine days since the long airlift anaesthetic and we hoped very much to spare him another long sleep.

'Be careful,' George warned Paul Sayer and I as we approached dressed in the rubber-lead protective aprons, 'he might not like the smell.'

George had a firm hand on Boy's neck and watched his reaction carefully. The odour of the aprons impregnated with countless animal scents did disturb him. Yet within a few minutes, during which I sat quietly next to him holding the X-ray plate in place under his leg, he succumbed to George's soothing voice; after a half-hearted attempt at chewing the plate edge, he gave a disdainful yawn and went back to sleep.

From beginning to end the whole procedure took two hours, the result being more or less what Toni had predicted: the T-fracture had diverted the pins from the direct path down through the centre of the humerus, and there was an enormous amount of callous formation at the fracture site. There were also a great number of small bone splinters which would eventually discharge themselves, mainly through movement as Boy took more exercise. The main thing the pictures did show was that the pins had held the bones for the initial healing period and that the broken ends had been correctly aligned. It seemed as if the pins would have to be removed in the very near future, possibly while we were away.

As soon as news of Boy's new location was voiced abroad a storm of controversy broke out once more over the lion rehabilitation project. Newsmen had found their inevitable way to George's prefabricated lakeside house and had taken innumerable photographs, which had once more stirred the public into action.

Many came forward, once again to defend George, among them Syd Downey, the famed East African hunter, and Major T. Norris, the naturalist. Emotion ran high and letters poured into the world press yet once again.

Throughout the storm Boy progressed, though another operation had to be performed, fixing a metal plate over the fractured ends of the bone which bowed outwards without the support of the pins. This slight movement would prevent final healing unless the bone ends could be completely immobilized for a period of time. Paul Sayer and an assistant this time performed the surgery while Toni and I, under the luxury of a roof and with X-ray at hand, administered the anaesthetic and supportive treatment which the complex surgical procedure demanded. In spite of the disappointment of another operation we reassured George that our original prognosis of at least 90 per cent recovery still held good. Boy was a young and basically very healthy lion. The bones had been aligned and normal function had already been demonstrated eight days after the 'bush' operation.

Again, Boy took it all in his stride. Not only did he make a remarkable recovery from the very long anaesthetic, but each time we visited him, handled and palpated him, he was perfectly behaved, although we suspected he would have been very different had George not been there to give him confidence. At times he became a little playful, pawing us, but always gently (judging by lion standards) and perhaps would take a nibble or two accompanied by a butt with his head. He always gave his welcome call, a mixture of purr and moan, whenever we entered his domain, and when we brought a stranger to see him he would make a point of including him, too, in the ceremony. It was remarkable how he continued, even with returning strength, to tolerate us, his tormentors. As he grew in health and strength he became more boisterous in his behaviour, until George himself found it difficult to keep his balance during the playful sallies, especially early in the morning.

At last, about two months after Boy arrived at Naivasha, he sent his first roar of dominance echoing across the lake. It was just about at dusk, George told us, when he began to pace up and down as if he was growing restless of his restricted environment. The discomfort of the second operation had worn off and his movement had greatly improved. He was no longer carrying the limb outwards, but more normally, and was beginning to sprint up and down the slope of his enclosure in sheer exuberance. Had he been a lioness, George said the ten-foot-high fence could have held him no longer. Even now it was becoming irksome and George was seriously – and hopefully – considering that the time had come to find him a new home.

Next door a tame young buffalo cow, which was herded with cattle, was becoming very much aware of Boy's presence. Every time he roared she bolted, and no doubt the entire bovine population of the Naivasha farming area within miles reacted to the fearful sound across the lake. As yet, Boy seemed content for most of the day, enjoying George's company, and his large daily ration of meat. As he became stronger he could no longer be trusted before midday; not because he had become vicious, but because no one, except George who was used to being knocked down in play, could withstand the force of those mighty paws or know how to counter his exuberance.

In April George left Boy in the care of his brother Terence, and began the search for new territory. He hoped to be able to settle along the coast north of Mombasa at Boni, but was thwarted by the rains which had already set in and had made the roads impassable. He returned to Naivasha and began again as soon as the rains had ended, settling on a remote region near Garissa with the aid of his old-time friend and fellow game-warden, Ken Smith. A highly experienced, now divisional warden, he pressed Boy's cause with enthusiasm and energy. He knew and had seen first hand what had been achieved with the seven rehabilitated lions in Meru Park. He respected and valued the contribution to conservation George had made and was determined to help him. He used his influence as a highly respected game authority to convince officialdom of the merit of George's work and did not cease in his efforts until a portion of the northern frontier area on the River Tana had been allocated to him.

'It's a good ten hours' drive on a very bad road,' George had told us. He had given us a little sketch map though we later discovered that he had not seriously expected us to find our way from this alone. He was due in Nairobi just about the time we finally had a free long weekend to go up, and right until we left we vainly hoped that we might be able to indulge in the rare luxury of driving up in convoy.

When by dawn of one cloudy wet Saturday morning there had been no sign of him we set off, five of us in two vehicles, complete with towing cable, shovels, safari equipment and food rations to last for two weeks at least. The short rains, calendar wise, were overdue; if the mountain streams filled the drifts and ditches and river beds, nothing but a helicopter could have moved us out.

A young camera-man who had already been up in George's new wilderness, filming the lion's release, had come to our rescue at the eleventh hour, giving us a fine, ornate map which served as an excellent guide. It showed different kinds of hurdles like fine dust and deep sand, long ruts, sandy river luggers, drifts which one came upon like elephant traps and trails wiped out by cattle hooves where one had to know west from east or get irretrievably lost. The map looked like something Boy Scouts used to issue for training in direction finding, except that all our clues were real and to pass the test was the only way to reach our destination.

The first 180 miles along the main Garissa road were passable but by no means without challenge. As we drove farther north and west it deteriorated, with rain clouds ready to break on to the corrugated slippery earth road which led to the Somaliland frontier. Even Toni, usually hopeful, began to despair for none of us could afford to take the chance of getting stuck in the mud for any length of time. George had told us rather casually, in true Kenya fashion, that we simply could not miss the all-important point at which we had to leave the main road and turn off north, for at the side stood a bush draped in toilet paper which was a permanent fixture.

'Some hunter put it there,' Terence explained to us later; 'not even the rain washes it away.'

True enough, the marker still held its flag of white drapes, and the road, from that point on, was miraculously dry just when we had expected the worst. At first we proceeded slowly, nervously, expecting a flooded mountain stream round every corner. After ten miles we began to relax and bless our good luck, savouring the scents of wild blossom which engulfed us along the lonely, winding, uncharted trail. We stopped and brewed coffee under the spreading shade of a tall acacia tree and listened to a crested lourie demonstrating at us from the tallest branch of the next tree with his 'go away, go awaaay' call. The lilac-breasted rollers flashed their rainbow-coloured wings at us, the Van der Decken's hornbills rocked back and forth upon their haunches in mating display. Suddenly, out of the denseness of the bush came a walking caravan of Orma tribesmen, their loaded camels like a mirage from the Arabian Nights. In contrast our vehicles seemed like decadent symbols of speed and civilization, noisy intruders of the wilderness we were invading. Yet without them we could never have come so far in such a short time – nor could return to our more mundane labour before three days were out.

I had never before penetrated this kind of trackless deserted country, except farther west and north, at Isiolo on a rhino-immobilizition exercise; but that had been long yellow grass, strong winds and vivid blue skies, and a horizon of rocks and clefts, so very different from the vast thorn-bush plains and distant palms.

We had covered forty-five miles from the turn-off when we found George's notice, proclaiming his land as Game Reserve. As we did, climbing to a steep stone-covered rise out of a long sand river bed, we saw waterbuck and baboons intermingling on the top of the incline. This was the first large group of animals we had seen, for so much had been hunted and poached out of existence; up to this point it had been only a flash of Grant's gazelle, or gerenuk or dik-dik in matrimonial pairs; the rest had been spoor of giraffe, elephant dung and many merging imprints in the mud and dust.

'When George manages to eliminate the poachers this is going to be another haven for the wild,' Toni said. 'What a marathon task, when every track and road has to be hand cleared; but he'll do it – with Terence to help him.'

If George would only be allowed to remain on after Boy and the other lions were completely rehabilitated, he could create a game reserve and eventually a Park as Colonel Stevenson Hamilton had created the Kruger National Park – 8,000 square miles as against 400 here, but another wildlife sanctuary just the same.

The last few miles to the new camp opened up vistas of rock and mountains we had not expected at all. As we drove uphill from the river the spectacular Kora rock range faced us, framed by dark storm clouds in the west. Someone had said there was a nice 'little hill' above the camp, so I expected something like Mugwongo. Kora, so named by the Galla tribe who used it as a meeting place before setting out for battle, was not by any means a 'little hill' but a high rock face which spread riverwards and into thick bush cover, the steep cliffs and the hump of one peak startling by its very starkness and contrast of light and white grey-browns against the many-coloured foliage into which it merged.

George, as we had feared, had not arrived, but Terence, his brother, was there to welcome us. I had never met him before for he lived and worked in the remotest parts of Kenya, carving settlements, bridges, camp sites and homes out of desolate bush and thick forest. Terence Adamson, senior to George, was a short, white-haired man, gentle-spoken, humorous, known to the Boran tribesmen as 'Abelone', father of cattle, who sited wells when all water failed and when the animals would otherwise perish of thirst. George needed his brother very badly at this time, for the creation of the new camp at short notice could not have been accomplished by anyone else. He was well known for his skill and his wonderful adventure stories of Kenya which he recounted with colourful detail as we swallowed pot after pot of hot tea.

'We've been expecting you for weeks,' he said, 'and now you've come George isn't here. If he doesn't turn up by dusk, we won't see him till tomorrow at earliest.'

In the late afternoon he took us down to the river bank and showed us where we could wash and cool off, warning us to keep a good eye out for crocodiles of which there was a multitude in the deep, wide palm-lined river. Clouds massing in the east could not dim the glory of that red tropical sunset, nor could the thought of angry territorial hippo or *Bilharzia* have lessened the cool bliss of the river. Refreshed and clean, we stood on the shore against the rocks in the fading twilight, unable to believe that we had really arrived in one piece with only two punctures between us after our long journey. If George did not come soon, for we could not stay beyond one day and two nights, we would miss him altogether.

He finally arrived that same night, two hours after Toni and I had dropped on to our bed in the back of the Wagoneer station-wagon. Suddenly there had been the flash of lights outside the fence and Toni's voice:

'It's George,' Toni said; 'I just knew he was coming tonight.'

I leapt up and out, shoeless in my pyjamas, and opened the gate to let the Land-Rover into the fenced enclosure. Toni was just behind me welcoming George, congratulating him on finding his way in the dark.

'But how did you find *your* way?' he said, smiling, looking rested and young instead of exhausted and strained. 'You're the first visitors who have come un-escorted. I must say, I am most surprised you got here at all.'

'Your map,' I teased him, 'and some old tyre tracks; and lots of luck, for we just *had* to get here, now, before the rains.'

Boy, looking massive and well, had arrived with Katania, the little orphan cub, and Christian, the lion flown out from England, rescued initially by two kindly Australians who had bought him at Harrods in London. He was a five-generations zoo-bred lion, fourteen months old, all legs, mane just sprouting, lumbering and very playful. George could not wait to greet them, nor they to greet him. They swarmed all over him, with that greeting moan I so well remembered, rubbing, nudging him, licking him, the cub Katania less boisterous, keeping her distance while Boy, checking Christian with wild growls every time he took liberties, continued his welcome for a very long time. At last, content that George was really back in their midst, Boy relaxed outside the gate and stayed there for the rest of the night, roaring triumphantly until the echo of his thunder shook the earth under our feet.

The clouds had become deep shadows, standing like rock-faces in the starlit sky, receding yet once again to build up, perhaps at dawn, and finally to break and soak the tinder-dry earth. George had gone to bed and I was just about to crawl under the net but found it hard to leave the recaptured night.

As I watched Boy, head on paws against which Katania, his special friend, rubbed her entrancing head, and Christian, bearlike and abandoned on his side against the fence, I thought again of George's immense struggle against the odds from the day he had found Boy badly injured, just over one year ago. Almost as soon as Boy had been released at Kora his limp had finally disappeared, for with exercise and movement the unused muscles had come back into play. Would Boy's fight back to health and freedom against so many odds convince the critics of the value of the experiment, whatever its outcome?

Chapter Twelve

THE ORPHANS

Not very far from where Boy was laying claim to new territory a young cheetah cub, barely two months old, had been deprived of its freedom. It was never known whether its mother had been killed or whether the cub had been taken from the lair while she was hunting. By the time it was discovered, starved, neglected and terrified in the hands of northern frontier bandits and brought to Nairobi by members of the Kenya police, it was hardly acceptable for its intended destination: an animal trader and finally a zoo.

Cheetah were more than ever at a premium as they became rarer, fetching several hundred pounds an animal. Their high value encouraged illicit trappers who shipped hundreds of cheetah out of Somaliland each year. In spite of new laws and a public outcry against the market in cheetah skins, the export continued and would continue as long as there was a demand. The numbers of cheetah were decreasing so rapidly that at last, they too, had been placed on the *red list* for endangered species.

At the time the bereft little creature came into our lives I was doing regular duty at the wild-animal orphanage which lies at the entrance to the Nairobi National Park, only eight miles from the centre of the city. I helped to care for the orphans, many of whom were in need not only of specialized veterinary treatment but also round-the-clock attention which could not easily be afforded them in the overcrowded, understaffed menagerie.

Samuel Ngethe, recently put in charge, and later sent to Bristol Zoo in England for further training, had asked me to come in for an extra visit before everyone went in different directions at the weekend: the ancient serval cat had suddenly died, probably of a heart attack, and Samuel wanted him checked externally before he was sent to the

College Veterinary Clinic for post-mortem. There was also a problem with the striped hyaena which had been receiving tranquillizing medicine for some time. She now no longer injured herself by trying to demolish her wire fence, but instead had dug herself into a hole from which she refused to emerge.

'Now, how to get her out,' he said: 'she seems so happy down there and only comes up to get food.' The hyaena was certainly a problem, but one which could hardly be solved on a Saturday afternoon. 'Perhaps she is tired of visitors,' I suggested, 'and feels she has done her share of entertainment. Why not just pension her off in a larger pen and let it go at that and perhaps one day someone will provide her with a mate?' The hopelessness of her state, since she was too old ever to be released even if she could re-adjust, impressed itself on me all over again. She was only being kept alive because she was a rare exhibit; if she remained hidden in her den she would, economically speaking, cease to earn her keep. Her days in the orphanage were obviously numbered.

The pygmy hippopotamus, apart from a cracked skin condition which had to be treated very carefully, posed a very unusual kind of problem. A gift to the people of Kenya from the President of Liberia, she had arrived as a female, but now, according to some opinions, seemed to have changed her sex. We had first met her at the annual agricultural show to which she had been sent immediately after arrival as a special exhibit. Confined in a most inadequate, unheated pool enclosure, torn from her West African climate to the cold high altitude of Nairobi, the displaced pygmy had become ill. Teetering on the edge of the concrete wall, Toni had injected her with a rhino-sized needle which had only just penetrated her thick leathery hide. As a result of this, or so we liked to think, she made an excellent recovery and was soon moved to her new quarters at the orphanage. I really had thought she was a female; had studied her from the tail down at close quarters for quite a long time. Like an elephant or a rhinoceros, her sex characteristics were not as simple to recognize as that of many other creatures who have been more simply and obviously designed – such as the lion or buffalo for instance.

Finally, after some embarrassing days of indecision, John Seago, the animal trapper, suggested that we consult Ted Reed, curator of the Smithsonian Washington Zoo, who happened to be in Nairobi and who had vast experience of pygmy hippos. 'I can see your problem,' he said kindly, 'especially as she has been sent over as a female. If she is ever to get a partner, we must make absolutely sure that she is really a she.'

'She' turned out to be a he. Ted Reed waited around for a very long time until the relevant portion of anatomy was displayed – and, after that, all doubt was dispelled.

'No woman vet ever knows how to sex an animal,' Toni commented unkindly, but nevertheless I noticed he was very relieved not to have been consulted directly on the matter, and that he had not, in fact, offered any advice on our quandary!

Brutus came next. He was an old lion, given to the orphanage by Carr Hartley, the Kenya trapper, who had kept him as a semi-pet for many years. He was getting cantankerous and difficult to handle but still required an enormous amount of food to maintain condition. He was a good zoo animal; massive with a thick golden mane. He seemed to enjoy the presence of visitors and would peer at them for hours while they gazed at him. He roared and rolled on his back in lazy abandon and sometimes even charged at the fence line when some arrogant visitor took liberties.

When he had first arrived Brutus was placed in the pen next to two Asian bears, who had been confiscated from a circus. He did not appear to be able to tolerate them, perhaps because of their un-African origin. We thought that the endless growls and aggressive display was only bravado to while away the time, but it seemed there was no hope even for a truce. One morning, during an especially fierce burst of temper at his neighbours, Brutus took a vicious swipe at one of the bears and must have caught his paw-pad on one of the metal strainers. By the time we arrived he was limping badly and holding up his foot, moaning softly as if asking for help. In the end Toni had to use a dart to put him out, after which we stitched his pad back into place.

Again, we made use of post-operative tranquillization and he responded extremely well. He allowed his wound to be sprayed daily as he drank his milk (which he enjoyed even in his old age) and after two weeks, when we had removed the stitches, he began to lose his limp.

'Come and see if he is fit to be moved to another pen away from the bears,' Samuel said. 'It would be terrible if it happened all over again.'

Toni and our friend Netta Pfeifer had come with me on this occasion. We drove our car down the inner orphanage lane between enclosures and parked outside the hospital opposite Brutus's pen.

He was lolling on his back, all four feet up in the air, apparently quite undisturbed by his audience of school children who tried to arouse him with pussy-cat enticements.

Samuel went to the larder and returned with a hunk of meat which he put into the far-end gate, away from the visitors. Brutus, though playful and relaxed at that moment, had immediately scented or perhaps seen the meat and loped over to take it, his movements free, his foreleg sound. As he passed the bears he gave a low growl and made a half-hearted charge, retreating before he reached the fence as if he had thought better of it.

'Move him as soon as you can,' Toni advised, 'before he injures something else. If you don't, double the wire of the fence in between.'

'Next week we will do it, when the labour returns,' Samuel said, 'but please now go and see the latest cheetah orphan. You might want to take him home with you,' Samuel added, flashing his winsome smile at me, from which I gathered at once that whoever the latest orphan was must be in serious trouble and needed much more than routine care. 'We will provide you with anything you need, a run, food, anything,' he added. Seeing my doubtful look, and making his excuses, he disappeared in the direction of the labour lines.

'Where are we going to put another animal?' Toni asked quizzically, 'especially one that needs complete confinement and careful watching. This one sounds as if it is going to be a handful.'

Our home had harboured all kinds of animal patients over the years, including young cheetah, but the proximity of the main road and neighbours' dogs had often given us second thoughts as to the wisdom of our home veterinary activities. Where on earth were we going to put a cheetah when we already housed a baby albino vervet monkey with a broken leg, a crow incapacitated by a broken wing, one bush squirrel inherited from a zoologist who had departed to England, quite apart from our houses guests, two children and two dogs?

'I'll be with you for a little while yet,' Netta said brightly, 'and I love cheetah; I'll look after it.' Netta, short, slim and puckish, was an American wild-life enthusiast who had worked for Joy Adamson for some months, helping not only with the cheetah-rehabilitation project, but with the book *The Spotted Sphinx*, of which she had typed the entire manuscript. She was a great favourite with Gail and Guy, then still at day school in Nairobi, and had often taken care of them and some of our animals whenever we needed a 'baby-sitter'. We wished we could have persuaded her to stay with us for longer, but her time in Kenya was drawing to a close – her visa almost expired. She felt it was time to get home to her dog and her family and she was going to see something of the world on her way.

'Marvellous that you're here to help,' I said to Toni. 'The cub sounds as if it might need sedation.'

Toni helped at the orphanage as much as he could, but had to fit in visits with his University duties. We performed surgery together here, as in the bush, when cases could not be moved to the College Veterinary Clinic. Anaesthetics, so different when applied to captive animals compared to free-living ones, were always in the experimental stage, especially when given to newly captured wild animals who were often injured and psychologically disturbed. From very light to heavy sedation, every chemical substance seemed to affect every kind of animal in a different way. Sex, season, age and health condition all played an important part. If wild animals were difficult to assess, then half-tame captive animals coming straight from the wild were infinitely more complex to subdue.

The orphanage was supposed to harbour sick, injured and orphaned animals in transit, keeping only a few on a permanent basis. This had been the founder's (Bobby Cade's) original concept; he had never intended anything even remotely like a zoo. Since he had died, three years after the orphanage had first come into being, ideas had changed and an increasing number of visitors had clamoured to see not only animals that came in for treatment and were moved out again, but others, which would remain permanently on display.

Already there were some permanents, such as Sebastian, the chimpanzee, who amused the crowds with his antics; there were the two Asian bears, the submerged striped hyaena; monkeys, jackal, cheetah, leopard and an assortment of antelope, often presented by people on leaving the country.

A new, larger orphanage was being planned, but would take a great deal of money and effort to complete. It would be a miniature zoo, a place where visitors could view animals at close quarters before entering the Nairobi Park, where they would see them roaming free.

We entered the little hospital equipped with several pens on each side, most of them occupied. A lesser flamingo had been brought in that morning, crumpled and inert, found on the shores of Lake Nakuru. Next to it was an injured grey duiker found on farmland, making a good recovery. There was a dik-dik female whose calf had died just after birth, and whose milk glands were full to bursting, in spite of being milked out three times a day. We had put out an SOS over the radio station for an orphan newborn small-type antelope and were still hopeful about getting a response. It seemed tragic that so much milk and mother love should go to waste.

The cheetah cub lay huddled in the straw of one of the large pens. As we entered he sat bolt upright, ruff up, spitting and growling as if he meant it.

'You might want to take him home with you,' Samuel had said; the question was going to be how to get hold of him in the first place. I approached very slowly, talking softly at each step at which he seemed to quieten down a little though the fierceness of his large, deep brown eyes belied his quietness. As I reached him he backed against the wall and glared at me, his eyes unwinkingly fixed on mine, rebellion, fear and suspicion emanating from the thin, sad little face.

I put out my hand to touch him, hoping to lift him up, but as soon as he saw the movement he became convulsed by another spasm of terror, as if not I, but my hand brought back some recent memory which spelt torture and a recall of his nightmare.

I turned to Toni and Netta, waiting in the doorway. 'Unless we catch him in a blanket, which would be very traumatic, or drug him heavily, which I would prefer not to do at present, there is only one way we can put him in a box and take him home: we will have to wait until he becomes used to us.'

Netta, her eyes sparkling with the challenge, moved a little way into the pen. 'Let me stay here with him while you do the rest of the rounds: perhaps I can tame him a little,' she said.

Toni and I had promised to go down to the Warden's house to inject an orphan lion cub against feline enteritis. It might take us half an hour or more since the Warden lived in the Park itself and the cub might not be so easy to restrain.

'The longer the better,' Netta said as she sat down in the straw. 'It is going to be a battle of his will against mine.'

When Toni and I returned half an hour later, Netta and

the cheetah had made friends. She was half sitting, half lying in the straw, the cub by her side, allowing itself to be stroked.

'He's purring,' Netta said triumphantly, 'but please don't come in now. Just leave the box inside the door.'

Toni went to get a top-lidded, straw-lined box and placed it just inside the pen. To our surprise we saw Netta pick up the little fire-ball by his scruff, put it in the box and shut the lid.

'Hypnosis,' Toni said; 'always knew you were a witch. What a pity you can't stay in Africa for longer and give this patient some psychiatric care.'

'He'll be fine,' Netta said. 'He just has a rebellious nature and needs lots of love. I would like to have him sleep with me tonight, if you don't mind.' Memories of Preech, our last little cheetah patient, who also shared his bedroom with a house guest, very much to the detriment of her beauty sleep.

We named him Zorba because of his fiery nature and his black moods, contrasting with his sudden volatile bursts of energy. He demanded attention, yet when it was given to him often resented it, like an orphan who yearns for love and cannot at first recognize it when it comes his way. With Netta devoting so much of the day to him he soon began to respond. He became very playful, though he would only play by himself, remaining aloof to any overtures made to him. It was a pity he needed therapeutic handling; his front paw had to be splinted; the back legs, rickety and twisted outwards, had to be strengthened with a support. We could not afford to wait to give him his treatment later, yet each time he was restrained forcibly a little of the ground that had been gained was lost.

We were testing the tranquillizer Valium on him but at first did not dare dose him too high. Being a cat, though some claim that the cheetah is more like a dog because of his non-retractile claws, he was bound to show some sort of reaction, especially as his nervousness and resistance to human contact would be certain to accentuate his intolerance of drugs.

To put it mildly, Zorba was a handful and we realized that if we were to make any headway with him at all, if we were to ready him for a life in captivity, if we were to instil some acceptance of man's dominance into the spotted spitfire, then he would have to have more hours of human company and attention than I could personally give each day, for Netta had most unfortunately left two weeks after Zorba arrived.

'Wish one or other of our angels would come and help us,' Toni said one morning. We had been most fortunate in the past, for a stream of delightful, intelligent female travellers who had been 'hooked' by the magnetism and beauty of East Africa had somehow found their way to our home. They had offered to work with our animals, help with research, typing or anything else that needed doing in return for their keep and the pleasure of staying in Africa. Toni had his own way of sizing up our female visitors, especially as he was always dreadfully pressed for help. 'When I meet a pretty girl,' he delighted in telling some of our more 'proper' guests, 'I don't believe in beating about the bush. I put the question to her straight out: "Can you type?"' There was something uncanny in the way Toni's wishes were so often granted; not in the misty distant future, but usually almost at once.

I should not, therefore, have been surprised when that morning, at coffee-break, Toni brought home a letter from an American friend, Cynthia Moss, who had assisted with elephant research at Manyara Park and had departed for her native New York via Naples, to visit her sister. While she was there, with only one more stage of her journey to go, she realized the intensity of her dislike of the life she was about to resume. In desperation she wrote to us asking if we could put her up in return for which she would help in any way she could.

'Just give me the word and I'll book my plane,' she wrote from Italy. 'I'm hoping so much that you'll have something for me to do.'

Something to do! I could not remember a time when there had been *so much* to do. Apart from animal patients at home and at the orphanage, last touches to add to my book which was due out a few months later, and Hatha Yoga classes, there was the family, house guests, and a constant stream of fascinating visitors to care for.

I wrote back the same day assuring Cynthia that she was most welcome and that I guaranteed to keep her busy for as long as she liked. I added a footnote with regard to Zorba, whom I knew would lure her to us faster than anything else. Even if the cheetah was not the tamest of pets, I was certain he would be so by the time Cynthia had given him care and love for a few weeks – or months as it turned out.

No one described our little spitfire better than Ralph Thompson, the famed animal artist, who came with his wife 'Bill' to visit us. I had always admired his superb wild-cat paintings, so much more alive than any other cat representation I had ever seen. Ralph's sketches and pastels were studies of wild-cat behaviour; he described the elusive, steel-spring ferocity of the leopard just as vividly as the lithe, playful grace of the spotted genet. He understood their personalities well, bringing them to vibrant life with pen and brush, so that they seemed to be about to spring at the beholder, no longer able to be contained by an inanimate frame.

We were having tea discussing their plans and ours, and the animals he yet hoped to sketch before leaving. We talked of Zorba and the other cats at the Animal Orphanage which might be of interest to him as subjects for his sketchbook.

'Excuse me for a few minutes,' Ralph said; 'I just want to take another look at the cub.'

Within ten minutes he returned, sketch pad in hand, and sat down to continue his tea.

'This is for you,' he said, handing me a sheet of drawing paper. 'It's poor but perhaps I can come back in a day or so and do one for my new book.'

It was marvellous! Ralph, in the space of a few minutes, had recreated Zorba with pen and ink, capturing his rebelliousness, his challenge and his every feature.

'No two alike, are there?' Ralph said, and how right he was. But to have caught that likeness, that untameable personality in a few strokes, was nothing short of genius.

By the time Cynthia arrived, two weeks after Zorba joined our menage, he had more or less become accustomed to his run, provided by the orphanage. He had begun to play with simple paper toys which I suspended from the wire roof which he attacked and demolished for hours at a time. He had put on weight and gloss and lost that haunted, hopeless look, tolerating (though with very bad grace) the double vaccination against cat flu. He growled and spat at our dogs and the monkey if they so much as neared the run, and even displayed at us if we approached too suddenly.

One companion he adored, even though he was torn and over-used: Gail's old pyjama dog which had been our previous cheetah cub's inseparable companion. I had almost decided to pension him off when Gail reminded me that he might well come in handy for another patient. Zorba loved the scruffy golden zip-fastener dog. Sometimes he romped with him quite gently, treating him like a lifeless toy: at other times he would attack him in real earnest, biting and pawing and tossing him as if his life depended upon it. Then he would suddenly stop, chewing and purring as if to make his peace with the rag doll, after which he stretched out on him with perfect contentment and went to sleep. Finally, after weeks of constant battering, even Gail had to admit that her pyjama dog had lost his shape and his purpose in life and that we had better look for another as soon as possible, since no veterinarian should ever be without one.

Cynthia, in contrast to the outgoing, urchin-like Netta, was quiet, soft spoken and introvert, her long flowing fair hair giving her the appearance of a teenager, so that many of our friends thought she was our other daughter. She worked with Zorba as soon as she arrived, spending hours at a time in the long, low run. As a good animal psychologist she knew that the best way to gain his trust was not to woo him too fiercely but to ignore him until, like a curious child, he would make the first approach. Cynthia – our cheetah girl, as Toni called her – would climb into the run complete with book and a soft webbing body-harness. Each day she sat with Zorba hour after hour, gently enticing him to play, or just sitting to keep him company. After a week, by which time he must have realized that his companion meant no harm, he began to romp near and even *on* her, biting her pages and pen in an effort to distract her away from her work.

Within two weeks Cynthia had become the centre of his universe though he still resented strangers in his pen. He was receiving daily small doses of tranquillizer but it was difficult to know how it affected him. As soon as he tolerated the harness we would have to consider returning him to the orphanage.

One thing was certain: we would not send him back until he had completely lost his fear of man and, moreover, his terror of *hands*. Even when he was at his most playful, and it seemed as if at last his reserve had broken down, the sight of a hand moving towards him would send him back into a nightmare of panic which sometimes took hours to dispel.

As Zorba grew in length and width as well as strength, so the fineness of his tear-marked face become more distinct and the baby fluff and darker underside of his body changed to a soft evenness of spots on light fawn. His expressive tail, long and black-ended, always gave a good indication of his innermost thoughts: he did not wag it with pleasure like a canine, but slowly, almost ominously lashed it from side to side whenever he was uncertain, angry or afraid. His fits of spitting and growling lessened quite suddenly, about the third month after he came to us; for the first time the main sound he made was a loud, reverberating purr, and sometimes he honoured us with that bird-like melodious chirrup, a sure sign that he was feeling more at ease.

Preech, our other cheetah cub patient, had chirruped right from the start so that, when we temporarily lost him in the garden, we had only to sit quietly for a few minutes and listen for his call which was like, yet more shrill than, a bird's. Rather the same as an immature distressed fledgling summoning its parents.

Manyaka (which means good luck in the Kikuyu language), our bush squirrel, already on arrival too old to be released, also chirruped for hours on end, emitting a less melodious chirp rather as if sandpaper had roughened his vocal chords. The injured crow – and we had several of them at different times – we kept in a cage next to his. Manyaka was the most sociable, welcoming animal, thriving on attention and company in return for which he would give an acrobatic display of the first order. Whenever the crow, who spent the days mostly in pacing his cage, came anywhere near him, the squirrel would somersault wildly, sending explosive bursts of sound across the garden even though he was upside down. The crow would watch the tumbling exhibition for some moments with deepest concentration, applauding the artist with a series of rasping 'caw-caws' as soon as he stopped. While Manyaka rested, panting but jubilant, the crow, whom Guy had named Cocoa, would sample the cheese stuck in the mesh of the squirrel's cage and then retreat into his own corner until Manyaka enticed him back again.

The mousebirds, blackheaded weavers and doves who resided in our garden were fed on the roof of the squirrel cage, quite unperturbed by either the crow who tried constantly to steal their food or by the antics of the squirrel beneath them. They appeared to communicate amongst themselves, squabbling over the food, and when it was all pecked up they flew away, leaving Manyaka to rest in the cotton-wool-lined nest of his inner chamber.

'How is Zorba getting on?' Samuel often asked when I went to the orphanage. Two other slightly older cheetah had arrived, a male and female from the same northern region, both fierce and unfriendly. I pleaded for another month so that I could increase Zorba's tranquillizing dose a little more, hoping to prevent any distress he might feel when moved from our home. Longer than that I could find no excuse to keep him. His legs had returned to normal, his furtive self-defensiveness had given way to self-confidence and a sense of mischief. His body weight was normal, his fur shining and healthy; from a jittery emaciated bundle of nerves he had grown into a truly beautiful, lithe spotted cat.

The day Zorba was due to return to the orphanage a gloom spread over our household. It was not so much his leaving that disturbed us, but the nature of his future home.

'I wonder if he remembers his cub days when he was wild,' Guy said at breakfast that morning; 'and if he does, if he misses them. Or if any of the orphans miss their real home.'

A sobering thought and yet Zorba, on that morning, was cheerful and tranquil, and did not in the least mind being placed into a straw-lined box, or travelling the eight-mile journey to his new home.

'Thank heavens for tranquillizers,' Cynthia said, amazed; 'he doesn't even seem to be put out by his unfriendly welcome.'

The female cheetah, larger than Zorba, who had been rescued one month before, did not respond to the new arrival's overtures. She arched her back, put down her ears and growled, long and deep, her eyes spitting anger in his direction. She was testy and suspicious not only of Zorba, but of everyone who approached her and no doubt she had every reason to be so, for the appalling state of her hind legs told a tale of neglect and illness.

Samuel threw Zorba a piece of meat but it did not reach its destination. With one lightning pounce, surprising considering her lameness, the dominant female snatched it off the ground before Zorba could collect himself.

'Please see that someone watches them when they are fed,' I begged Samuel. Poor Zorba, he was in for a very rough time which could hardly be compensated for by the company of his own kind.

We could not bear to stay and watch any longer though we knew that eventually, within two weeks as it turned out, Zorba, the female and the other male cub would make their peace. At that moment when we took our farewell I felt as if we had deserted a friend and that we should have somehow spirited him away into the wilderness where he truly belonged. Unreasonable, irrational, perhaps over-sentimental, for Zorba could never have survived alone. Yet for days his face kept impinging itself before my eyes; perhaps not so much his face, but those eyes: unflinching, defiant, independent, restless cat eyes, the call of the wild unmistakably beaming out of them so strongly that no tranquillizer in the world, nor any amount of taming, could have changed it.

Chapter Thirteen

HUGO MY FRIEND

Not only Hugo's eyes, but his face and demeanour were tragic, perhaps the most sorrowful expression I have ever seen. He was a young chimp, we never knew exactly how young, but guessed between twelve and fifteen months; by chimpanzee standards early infancy.

Immobile, as if stunned, he was waiting for us on the lounge coffee-table in a wooden box, lid ajar. The black-pink body was quite inert, all but the expressive eyes which opened wide as we entered, pleading, yet frighteningly resigned as if he was aware that life had passed him by and that as a homeless refugee he had no right to expect anything good to come his way, least of all a grain of happiness. I untangled him from his box very gently and took him in my arms, letting him cling to me while I rocked him gently back and forth, trying to convey to him that he was safe.

'Ug, uuugh, ug, uuu uuuugh': he vocalized softly, lips closed, his belly vibrating slightly with each sound against my chest. I nickered in response, stroking his back, horrified to feel very little but bone from neck to coccyx.

'He is a very ill, stressed chimp,' I said, as Toni came back into the room with a thermometer. At that the tightly intertwined arms slackened a little and the large head leaned back, the deep glistening brown eyes assessing me as he pouted his lower lip for a moment as if to test the honesty of my affection.

'What did you tell the owners?' I asked as Toni placed the thermometer under the chimp's armpit and steadied the arm to keep it in place.

'I didn't tell them anything except that we would take a look at him first before giving an opinion. They are coming back in an hour for our verdict.'

Even before we had factual evidence of his fever I knew

we could not let him go. There was shock and stress, a bladder and intestinal infection as well as parasitism and malnutrition. He had travelled in a most inadequate box and had probably been cramped into it for a very long time, judging from his complete inability to extend either his limbs or body. His health papers, we were told, had expired by the time he arrived, from which we concluded that there had been a delay in shipping him from the Congo. He had been caught with 70 others by a commercial trapper but we could not discover anything else of his history: whether his mother had been killed or had died in captivity, and for how long the chimp had been held before she had died, if she did. What kind of life had he been subjected to that had turned him into a half-alive, frightened, stunted animal?

Toni and I had never looked after a chimpanzee before, other than the occasional case at the orphanage, when Sisi or Sebastian needed behind-bar treatment. How fortunate that we had visited the chimpanzee Reserve not so long before and had at least gleaned a little of chimp behaviour. We knew, for instance, that to gesticulate animatedly with one's hands and arms was taken as an aggressive movement and that the lifting of the lips, our smile expression, meant recognition and pleasure. We knew the need for the clinging comfort that the baby chimp derives from close contact, though neither of us were prepared to have our newest patient cling to our stomachs indefinitely. We had seen and marvelled at the wild chimps' agility in voice and movement and hoped that Hugo, named in honour of our host in Chimpland (Baron Hugo van Lawick), would soon become active and vocal too.

By the time the owners returned, long after dark, we had made up our minds to keep him until he was really well, however long that might take. To make him feel at home (home at night being a nest high up on a branch), Toni had suspended a straw-padded kikapu – locally woven basket – by a long rope to a nail he drove behind the picture rail. Though long past his bedtime, we thought it best to keep Hugo with us while he was so sick, enticing him with pawpaw after he had accepted his first dose of medicine from a spoon without protest.

Nairobi was very chilly compared to his native Congo, so we wrapped him into a blanket as soon as he had drunk a cup of yoghurt, the latter a marvellous stand-by in cases of debilitation and upset digestion when the patient cannot tolerate or digest rich food.

Hugo settled into his basket very happily, dozing and waking in turn, craning his head over the rim every now and then to make sure we were still there. He said very little that night, falling into an exhausted sleep from which he did not wake even when we moved him into our bedroom, suspending him, this time, from an open wardrobe door. 'Hope he comes to life soon,' I said, checking Hugo for the last time.

I wondered how long it would take for him to combat the infection and overcome the effect of his flight which must have been terrifying. Would he ever be able to stand up straight, like Figin or Humphrey, the magnificent males of the Gombe troop? Or vocalize in that irrepressive, joyful way, or swing up into the trees, as if gravity were an illusion? Just as I was getting into bed Hugo gave his first loud call – short and bird-like – and I felt guilty that we had disturbed his precious sleep. But when I looked at him his eyes were still tight shut, his hands gripping the little blanket under his chin like a baby secure in his crib.

Five days later we found it hard to believe that he had *ever* sat still during mealtime, or at any other time for that matter!

He was given a special chair next to Toni, and we tried to impress on him that this was his place and his alone, just as our chairs belonged to us. For the first few days this worked extremely well and any friends who shared the table were impressed beyond measure at our patient's impeccable table manners. Had they returned at the end of one week they would have seen a very differently behaved ape, for Hugo was fast regaining energy and high spirits and an almost miraculous propensity for being where at the same time. He firmly believed that what was ours was his also and that he had every right to register violent protest whenever we broke his rules.

Hugo was an early riser, climbing on our bed and waking us gently with first light, grooming Toni's beard lovingly with his long black-nailed old-man fingers. When we stirred he was ecstatic, vocalizing cheerfully until we got up. By then Toni had purchased a baby-blue jersey to keep off the damp night chill, though it was anything but baby-blue by the morning. While we washed, dressed and exercised, which was the time Mr Seb also joined us, Hugo would be busy assessing our room contents, sampling the remains of the morning lemon-water, from which he extracted the slice of lemon for a breakfast snack.

While I stood on my head, a posture Mr Seb always watched from a safe distance since the time I had come crashing down on to his head, Hugo tried his best to disturb my balance by climbing on to my stomach – the only time I could not resist. For the first two weeks Hugo took his food ravenously, especially at breakfast time when he swallowed his bowl of fruit, yoghurt and cereal with incredible speed. In spite of his huge appetite he always preceded his repast with a flow of appreciative sounds during which he kept a very close watch on Mr Seb and Jessie in case they had any ideas about sharing his food. At such times he always turned up his lips and smiled, or rather grimaced, conveying that he was pleased and perhaps even grateful. If ever Jessie took one step forward towards his dish Mr Seb intervened on Hugo's behalf with an ominous growl, a deterrent which Jessie had the good sense to heed. Even Hugo, though an invalid guest, had to learn to respect other's food the hard way. One lunch-time he took the liberty of dipping his long fingers into Mr Seb's bowl while the latter was eating. We were only aware of a growl and a scream and Hugo's sudden catapulting movement as he leapt on to Toni's chair, his frightened face buried against Toni's back. With one hand he firmly gripped Toni's waist, secure in his citadel, certain of protection. With the other he first tested his slightly bloodied nose, then flung it towards Mr Seb

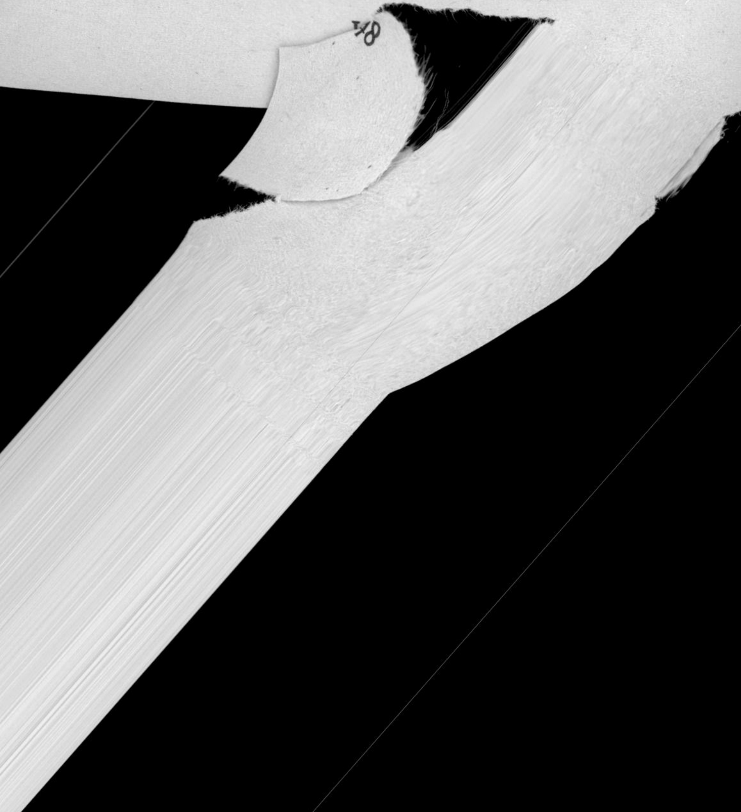

him with pawpaw after he had accepted his first dose of medicine from a spoon without protest.

Nairobi was very chilly compared to his native Congo, so we wrapped him into a blanket as soon as he had drunk a cup of yoghurt, the latter a marvellous stand-by in cases of debilitation and upset digestion when the patient cannot tolerate or digest rich food.

Hugo settled into his basket very happily, dozing and waking in turn, craning his head over the rim every now and then to make sure we were still there. He said very little that night, falling into an exhausted sleep from which he did not wake even when we moved him into our bedroom, suspending him, this time, from an open wardrobe door. 'Hope he comes to life soon,' I said, checking Hugo for the last time.

I wondered how long it would take for him to combat the infection and overcome the effect of his flight which must have been terrifying. Would he ever be able to stand up straight, like Figin or Humphrey, the magnificent males of the Gombe troop? Or vocalize in that irrepressive, joyful way, or swing up into the trees, as if gravity were an illusion? Just as I was getting into bed Hugo gave his first loud call – short and bird-like – and I felt guilty that we had disturbed his precious sleep. But when I looked at him his eyes were still tight shut, his hands gripping the little blanket under his chin like a baby secure in his crib.

Five days later we found it hard to believe that he had *ever* sat still during mealtime, or at any other time for that matter!

He was given a special chair next to Toni, and we tried to impress on him that this was his place and his alone, just as our chairs belonged to us. For the first few days this worked extremely well and any friends who shared the table were impressed beyond measure at our patient's impeccable table manners. Had they returned at the end of one week they would have seen a very differently behaved ape, for Hugo was fast regaining energy and high spirits and an almost miraculous propensity for being [illegible]where at the same time. He firmly believed that what was ours was his also and that he had every right to register violent protest whenever we broke his rules.

Hugo was an early riser, climbing on our bed and waking us gently with first light, grooming Toni's beard lovingly with his long black-nailed old-man fingers. When we stirred he was ecstatic, vocalizing cheerfully until we got up. By then Toni had purchased a baby-blue jersey to keep off the damp night chill, though it was anything but baby-blue by the morning. While we washed, dressed and exercised, which was the time Mr Seb also joined us, Hugo would be busy assessing our room contents, sampling the remains of the morning lemon-water, from which he extracted the slice of lemon for a breakfast snack.

While I stood on my head, a posture Mr Seb always watched from a safe distance since the time I had come crashing down on to his head, Hugo tried his best to disturb my balance by climbing on to my stomach – the only time I could not resist. For the first two weeks Hugo took his food ravenously, especially at breakfast time when he swallowed his bowl of fruit, yoghurt and cereal with incredible speed. In spite of his huge appetite he always preceded his repast with a flow of appreciative sounds during which he kept a very close watch on Mr Seb and Jessie in case they had any ideas about sharing his food. At such times he always turned up his lips and smiled, or rather grimaced, conveying that he was pleased and perhaps even grateful. If ever Jessie took one step forward towards his dish Mr Seb intervened on Hugo's behalf with an ominous growl, a deterrent which Jessie had the good sense to heed. Even Hugo, though an invalid guest, had to learn to respect other's food the hard way. One lunch-time he took the liberty of dipping his long fingers into Mr Seb's bowl while the latter was eating. We were only aware of a growl and a scream and Hugo's sudden catapulting movement as he leapt on to Toni's chair, his frightened face buried against Toni's back. With one hand he firmly gripped Toni's waist, secure in his citadel, certain of protection. With the other he first tested his slightly bloodied nose, then flung it towards Mr Se[illegible] again and again with the gesture of a furious child, hurling a flow of abuse at him for almost fifteen minutes. Our Alsatian, having emptied his bowl, lay down and faced the chimp with icy reserve, listening to the flow of ungentlemanly language with his fawn-coloured eyebrows slightly raised.

The week after Hugo was given to our care, Bambi, a young bush-duiker arrived, very severely injured, yet just as friendly and a delightful patient. She had been found by David Sheldrick, the Warden of Tsavo East National Park, tied and strung up at the local shop in Voi Village – live meat on the hook as it were – for the highest bidder. David had taken her home to his wife Daphne who had tried to nurse her back to health, an almost impossible task when there was already so much damage to the bones. Infection set into all the joints and by the time Bambi reached us she could not stand at all, though she tried to propel herself on folded legs.

Mr Seb, perhaps stirred to pity by the prone antelope, this time took complete possession, resenting not only Jessie, but any visitor who came nearer to Bambi than he thought fit. He would get up and lick her with gentle affection to which she responded like a lady in love, touching noses and bleating gently. Mr Seb would not rest until the 'danger' had passed, when he would again settle down next to her, very much on guard.

The chimp paid little attention to Bambi though sometimes I saw him sitting with her while he groomed himself or used her as a pillow while he dozed in the sun. In the same way she took little notice of him, accepting him as part of the landscape, like the birds and lizards. Their first meeting had been neither affectionate nor dramatic, yet the relationship between Jessie and Mr Seb and the two wild ones was anything but indifferent. The communication and fellowship between our tame and wild animals was so constant that it could, no longer, be taken as coincidence; the two canines accepted our patients as guests who needed care even if, genetically speaking, they should have regarded them as prey. Was fear and aggression, after all, an environmental rather than an inherited factor?

Very soon, compelled by the change in behaviour as Hugo regained his health, we decided to let him have his own, private, night 'nest', the little wooden house at the end of the garden attached to a wired-in run. He had grown very much stronger and demanding, creating havoc in our bedroom if we left him alone. He slept less heavily and would get up at night to walk about like a restless spirit, waking us with his jubilant bird call which seemed to express new-found emotion and pent-up energy. If we stirred or conversed or put on a light he no longer slept through it but came to life at once, head up over the rim of the basket, protesting loudly and shrilly at the interruption of his rest. It might be less cosy for him in the garden, but it would be more restful for all of us. The time had definitely come for Hugo to sleep alone. The first time I took him down, just at dark, half a pineapple in hand, I thought dreadful screams would rend our neighbourhood. But nothing of the sort occurred. He inspected his new quarters of box and straw, dived in after his pineapple and was heard of no more until the morning.

Each day after breakfast Hugo had to be scrubbed, a process he tolerated and even enjoyed, especially the rub-down afterwards. He uuuughed contentedly as I brushed his nails and teeth and did not even object to having his bottom washed. As long as he was getting attention he was content; when we had visitors his bliss knew no bounds, especially when children came to see him and play with him.

For *at least* one hour of the day Hugo needed to cling, best of all entwined round a neck, second best round a leg or chest. Third choice was to sit in my desk-drawer chewing raisins or dried apricots or yeast tablets one by one while my typewriter clicked away. If he got bored with that, I gave him the waste-paper basket full of paper which he investigated with great care, piece by piece, testing each item to see if it really was useless, gastronomically speaking. After that he would try to attract my attention with his antics or just sit and think and talk to himself; or else swing on to my legs to nap until there was something better to do.

Toni loved Hugo; I had never before seen him fuss so much over any single animal. When we ate our meals he complained terribly if Hugo did not keep his hands off the table, yet often ended up by sharing his supper with him while I remonstrated in vain about over-indulgent fathers!

'Hugo, my friend,' Toni would say as he gave him his second helping of fruit salad, 'Hugo, you and I have a strong link – we understand each other.'

Hugo's tastes were fairly consistent; he adored pineapple and pawpaw, home-made wheat bread, yoghurt and all cereals; he was bored with bananas, probably the only diet he had received after capture, ate avocados if they were not over-ripe and any dried fruit including dates. Though at first he had refused the bottle, seeming almost too weak to suck, he finally worked up enthusiasm and took the teat, eyes closed in ecstasy, as if he was making up for lost time.

Other than his normal diet he ate *anything* that happened to be on *our* plates, in *our* cups or in any way classifiable as *our* food, no matter what the taste. If I came towards him with a spoon of medicine he opened his mouth into an almost beak-like shape of welcome, swallowing dutifully even if the contents were not perhaps as delicious as he had hoped. His habit of turning his head away, hiding his face and wrapping his long arms around himself while his long-toed feet stood on each other, disappeared altogether as he learnt to trust us. He sometimes still half-closed his eyes as if to shut out what he was seeing, which might be the end of a dog-lead and rope which I sometimes had to fasten to his collar when I left him in the garden. He would then show his resentment by drumming his hands on the ground and screaming with temper; when I went out to reason with him he stopped at once, ashamed at having caused such an uproar, grooming himself as he always did when he was embarrassed.

He loved beards, buttons and pocket-books and George Adamson's cravat which he tried to tease out of its knot to the accompaniment of popping, lip-smacking grooming sounds.

What worried us was that he was quite unable to play, as we had seen young chimpanzees play in Gombe Stream. We had heard them laugh and chatter as they swung up and down, chasing each other from tree to tree, their immense sense of fun and agility two very marked features of their species.

Hugo, snatched from his homeland, had missed the playtime of his life, for he had probably lost his mother at a time when most baby chimps begin to enjoy life. We tried to teach him to play and Jessie tried to entice him with her endless ball games; but Hugo had no idea what he was supposed to do; it was sad beyond measure to see an ape without a sense of fun.

'. . . wild chimpanzee offspring, just like human children, love to play with objects as well as each other,' wrote Jane van Lawick-Goodall.* 'Their forest home provides some wonderful play things . . . Young chimpanzees in the wild are quite as playful as they are in captivity.'

He also had no wish to climb trees, which in a tree-climbing species seemed sad beyond measure. Perhaps, later on, he would become both agile and playful, but now we were thankful that he had recovered his health, had put on four pounds in three weeks and had lost the 'pot' belly and 'razor' back appearance.

It was feared that he might turn vicious as male chimpanzees usually do as soon as they reach sexual maturity; if this happened his life among people who cared might not be very many years long. He seemed so incredibly human that one's responsibility to him was so much the greater, almost as to a child one has adopted. If he could be released back into the wild all might yet be well for him, though there was always the fear that wild territorial chimps might tear him to pieces. Like all other orphans displaced by human intervention, his future was a large question-mark.

* *National Geographic*, December 1965.

Toni and I felt a kind of personal loss when Hugo departed to his new home, rather as if a member of the family had gone. Suddenly there seemed so little to do; the verandah was tidy again, the large monstera plant stood up straight instead of hanging at a downward angle. No more flower pots were broken, no chairs upturned, no pieces of discarded pulp-chewed pineapple or pawpaw on the lawn.

'Where is Sokomotu?' asked Paulo, our gardener, who had brought his wife to call.

'Where is Hugo?' asked our friends, and especially our friends' children.

'He is well now,' I explained to them. 'He has gone home but he will often come to visit us. When a patient needs no more care from his doctor then he must go where he belongs. There is sure to be someone waiting to take his place.'

Our Sokomotu which is Swahili for chimpanzee, but literally translated means the market man, who sings his wares from the tree-tops, sang in our garden no more, at least not for a little while. As we had predicted, he came back to stay as our house guest when his owners were away, adjusting from one menage to another more easily than a seasoned traveller. He needed so little in life and travelled lightly, his only worldly possession one baby-blue woollen jersey.

Chapter Fourteen

FULL HOUSE

Bibi Kali (the fierce lady) was neither as long nor as broad as my hand, and lay so still that I thought she was dead. A letter from Samuel Ngethe, of the wild-animal orphanage at the entrance to Nairobi National Park, lay beside her:

'Dear Doctor Sue, please stay with this leopard cub until she is well. She was born today, and hurt by her mother.'

I picked up the inert, furry body and brought it into the light; there *was* a regular slight movement of the flanks and when I held her against my face I could feel the minutest flow of warm air against my cheeks. The eyes as always at birth, were tightly sealed and would stay so for at least a week. I slapped her gently, and rubbed her, getting in response a flicker of movement in the hind legs. She certainly was very newly born, judging from the condition of the umbilical cord which hung down soft and damp from her belly. The nasty-looking tear wound across the shoulders would have to wait.

I tucked her into my bosom, put on another jersey to ward off the chilly night, and went to find dried milk and a doll's feeding bottle. Whatever I gave her, I could never replace the rich, stimulating mother's first milk, the colostrum, with its high antibody content. By the time the bottle was ready the cub felt a little warmer, but did not yet have the strength to suck. I found an eye-dropper and forced the milk down drop by drop, trying to establish the swallowing reflex by softly massaging her throat and alternately closing the nostrils, so that she would have to open her tightly clenched jaws to get air.

After a few moments of struggle she began to get the idea, allowing the milk to flow into her mouth from the back corner of the jaw to prevent choking. Suddenly, as if the consciousness of life had entered her body at last, she gave a surprisingly powerful wriggle of her back legs and began to cry, the sound identical to that of a baby.

'She wants more!' Toni, who had come to watch, filled the dropper for me and this time included a tiny amount of Abidec, the multivitamin infant mixture. She seemed to like the flavour, eagerly swallowing and craning her neck forward to get at the source. After the third dropperful she relaxed and made no more sound. The little cub had returned to sleep.

I wrapped her up in a baby blanket, placed her on to a hot-water bottle and put a ticking clock under it all to act as a surrogate mother's heartbeat, essential if a newborn orphan was to thrive. The sounds, first in utero (in the womb before birth) and then after birth when the young remains close to its mother, are always there; movement sounds, heart and breathing rhythm, the gurgling of the digestive tract. After that clamour, the sudden withdrawal of the mother presence, as well as scent, tongue massage, vocal communication and teat texture may cause stress and shock so great as to result in death, especially in animals which live very close to their mother for weeks or months after birth.

'Looks as if she might live,' Toni said.

Was her mother wild or captive and why had she injured her young? It is known that cats, wild and domestic, sometimes eat their new born; yet causing them injury, except by accident, is most unusual. One could only marvel at the leopard's continued survival in the light of the onslaught by pelt hunters and poachers and wonder how much longer they can hold out. How does the leopard thrive while man increasingly curtails his territory?

I looked down at the limp mass of cold skin, spotted fur, drooping tail and unresponsive limbs and remembered the large, green-eyed leopard I had seen not so long before in the branch of a wild fig tree in the Serengeti National Park, in Tanzania. I recalled the powerful steel-grip jaws, supple back, black rosettes on gold brown fur and the long, expressive tail, solid wide paws with needle-sharp claws which grip and rip with terrible ferocity. The shadows of the tree hid much of his lazily stretched body, yet from his eyes and ears I could detect his alertness, his lithe feline dominance. Just such a leopard had been seen near a 200lb. infant giraffe carcase which he had dragged twelve feet up on to a branch. The leopard had weighed 120lb. at most, yet somehow he had found a way to hold and drag nearly twice his own body-weight into the tree.

He seems too secretive, too cunning a creature to allow much to be discovered about him. He has been studied to a small extent and some information has been gleaned from those that are trapped, marked and released. Yet it is this very cloak of mystery, the wide variety of his diet and his adaptability which, at present, seem to be protecting him from eradication. Even so, apart from man, he does have natural enemies: hyaena, crocodile and lion who, contrary to belief, will steal the leopard's kill out of a tree whenever they can.

'This cub was born in the orphanage,' Samuel told me the next morning. 'The mother has been with us for some years, and she has never injured her young before. Someone said the little one went too near the fence and she tried to pull it back.'

It was also suggested that she had been upset because the male leopard had remained in the enclosure with her while she gave birth. In the wild, the male seems to make itself scarce at such a time, as does the lion and the cheetah. The leopardess has to live by her wits and her wits alone, depending on no one; if the natural habit is upset at times of crises, such as birth, the consequences can be disastrous.

It was three days before we dared to give anything but superficial treatment to the shoulder wound. By then the cub was taking milk from a bottle and was making her presence felt in no uncertain terms. She seemed gradually to take shape, coming to life with an infant bawl rather than a feline cry, her body, now well nourished, taking on the supple rippling movement of the adult leopard. We had overcome the constipation by emulating the mother's warm moist tongue with warm cotton wool on the appropriate part of her anatomy several times a day. Once that problem was solved, she began to thrive and so we cleaned up her wound, injected a little local anaesthetic and stitched her up, the real problem not being the actual surgery but how to keep her still while we performed it! In spite of the lag in time between infliction of the tear and our repair, everything healed up perfectly, so that when we removed the stitches after seven days, there was hardly a mark, except where we had clipped away the hair.

'We must try and get a foster mother for Bibi Kali,' I declared one morning; 'I wonder if we should advertise or just ask around among our friends. It would have to be a bitch with all her primitive instincts bred out of her, otherwise she might react just like Rolf.'

'Why a dog?' Toni asked, a little sceptically; 'surely you can't mix felines and canines. If you do, dreadful things might follow and the leopard might eventually eat the mother.'

I reminded him of Za Za, my miniature poodle patient, whom I had fostered on to my generous domestic cat. The pup's personality seemed to have been unaffected by her mixed upbringing except that she tried hard to mew when competing with the rest of the litter to get at the milk. Apart from that she had thrived and was now the revered great-grandmother of many pedigree miniature poodles, who seemed as normal as any of that highly strung, un-canine breed.

That teatime, as if in answer to a prayer, Kay Bennet came to see us with her little Maltese bitch Minky and a two-week-old pup which was looking for a home. Being the mother of a game warden – Ted Goss of Tsavo West National Park – and in any case one of those remarkable women who are keen to try anything crazy at any time, she agreed to give it a trial, promising to bring Minky to us the next day complete with diet sheet, basket and puppy.

'She has quite a bit of milk, but then her pup has a two-week start. I hope the leopard will get a look in.'

Minky was perfect for the job; she was placid, very easy to handle and had plenty of milk. She was an ultra-sophisticated city dog whose line had long since lost any primitive fear of leopard. Her pup, his eyes already open, was completely unconcerned by his new spotted sister. As long as there was enough for him to drink, he didn't mind if anyone else took the left-overs.

Or so it was at first! Bibi Kali was just opening her eyes the day Kay brought Minky to us, the sixth day after birth; she had to be coaxed while I guided the leopard cub to the distended teat and held her down while the cub suckled. But only for the first twenty-four hours. After that she settled down to nursing her two babies, resigned to the situation which brought her a great deal of praise and attention as well as additional food.

On the tenth day I noticed that the cub was beginning to push the puppy from its teat and that Minky did nothing to help when her first-born cried with frustration and anger. By the fourteenth day I decided to supplement Bibi Kali's milk by giving three bottle feeds as well. That meant that the puppy, though still twice the size of the leopard, was no longer having to compete at some of its meals and so everything went much more smoothly. By the eighteenth day, having tripled its size, the cub began to do some serious treading and pushing against Minky's belly as she drank, holding on to the teat for dear life. Minky, as if suddenly aware that her foundling had retractile claws which tugged painfully at her flesh every time it clung on, did not quite know how to handle the situation, issuing pathetic little cries whenever we passed her basket.

It was Toni who first discovered the cause of her trouble, and suggested that we clip the cub's needle-sharp claws. This worked like a charm. The cub objected, quite naturally, and growled at us for the first time, the sound deep and rasping, so very different to the infant cry we had become accustomed to. Her new-found voice and increasing agility were fast transforming her into a young leopardess and I began to wonder whether Toni's prediction had not, after all, been correct. If our fierce little vixen developed her leopard temperament as quickly as she had developed her growl, she might well make mincemeat of her adopted brother in the near future.

'She might even finish up by making mincemeat of Minky,' Toni warned again. 'You had better wean her to solid food.'

There was a lot in what Toni had said and I thought it best to heed his warning. Kay Bennet was devoted to Minky and would never forgive us if she was devoured! I began to feed frequent small amounts of minced liver to the cub which it ate ravenously, growling fiercely all the time as if attacking its prey. This, I hoped, would allay any carnivorous intentions the leopard might have towards its adopted family, until such time as her foster mother's milk dried up and she left for home.

From this time on the puppy and leopard began to enjoy very lively games, rolling and tumbling out of the basket which hitherto had acted as a barrier between them and the outside world. That world, they now discovered, was so large that they could get lost in it without any trouble at all. Had they been silent while they frolicked and disappeared under cupboards and down the verandah steps I think most of the day would have been spent in trying to find them. As it was they vocalized, each in his different language, to which Minky sometimes added her anxious yapping in case the horseplay got too rough. On occasion she actually pursued them and joined in, her pup's and her own white silky coat contrasting strangely with the leopard's very pronounced spotted fur and sturdy long-tailed body. In contrast the two canine aristocrats suffered from the indignity of docked stumps, which hampered their self-expression enormously.

Thank heavens Cynthia Moss was with us at the time, to help with our swelling wild feline population, for we already had a genet female we had reared, known as Poppet, who admittedly slept all day but demanded a great deal of attention at night when she woke up. She had grown supple and very playful, embarking on nocturnal games with Jessie whom she teased mercilessly by the hour. She was so lithe and fast in her movements that by the time

Jessie realized who it was that had given her a sharp nip, Poppet had leaped away and out of reach, laughing down at her from the height of a chair or pelmet.

As Bibi Kali grew in size and in speed so she attempted to join in the games, using the genet's long tail as her exclusive toy.

The only one who could not take part was the sad serval cat, who had recently joined us from the orphanage. She lay in her basket with her ears pricked, watching our activities with great interest. Whenever I approached to feed her she mewed plaintively and tried to sit up, threshing with her front legs until she became exhausted. Ralph Thompson had sketched her and her three kittens a few days before he had left for England and had remarked on the special beauty of her black on deep yellow markings and her long sensitive ears and brown-green grey eyes, which spoke of great intelligence.

She had demonstrated this amply at the time I had to vaccinate both her and her kittens against feline enteritis, the deadly disease which had already once swept through the orphanage leaving a great number of casualties in its wake. I had thought that the wilder of the cheetah would be the most difficult to inject, yet once lured into a box-trap with food they posed no further problem. In the end it was not the large, but the smaller of the African wild cats such as serval and caracal that showed most resistance, resenting our strange behaviour and the disruption of their routine to the last.

One of the animal scouts had gone into the serval pen with a catching net at the end of a stick but found his quarry very elusive. The kittens, fast as lightning, appeared to play a game with him, most ably assisted by their mother who guided them into the cramped space behind their house where it was almost impossible to reach them. Finally the scout captured one kitten under his net, calling to me to come in and inject it. He had released the long stick and was turning his attention to the other kittens when the serval mother leapt towards her young, and picking up the metal rim of the net in her mouth, released the captive. While the scout, Samuel and I watched in amazement at this display of animal intelligence, she pulled the net towards the back of their sleeping quarters, frustrated in her efforts when the long stick was caught against the side of the fence.

No one ever knew what had caused the fracture of the serval's spine; it had happened quite suddenly for on the morning before I came to see her the scouts reported that she was well. She had been moved into the straw of her sleeping house and was lying on her side, her kittens huddled into the opposite corner, frightened and shivering in the rain. While I examined her she made several different kinds of mewing sounds, the longest, softest, no doubt intended for her kittens, the deeper, louder sound, I thought, intended for me whenever I touched a sensitive spot. She would then, very gently, grip my hand with her teeth, letting go only after I had stopped palpating her. I could not feel an obvious break, yet could not take the chance. If she had received a blow on her back with consequent bruising and pressure on the spinal nerves, the symptoms, at first sight might be the same.

'What about her kittens, if I take her away?' I asked Samuel Ngethe. 'They are terribly wild and will need great care in handling and feeding. If we can find someone it might be better if they do not stay with their mother, for it distresses her terribly that she cannot look after them.'

'I will ask the Warden. He has an enclosure below his house; perhaps he will take them.'

When I went to visit the kittens in their new home next to the Warden's house one day later, they seemed reasonably content and were receiving the best of care. They missed their mother, of course, and we hoped that she would be restored to them, in time. If not, they could grow up without her.

But the serval with the shining eyes and fine intelligence was doomed. After a week at our home, during which she lay peacefully in a Moses-basket cradle but made no progress, we took her to the veterinary clinic to be X-rayed. When the result came through it was found that her spine, two-thirds of the way down her back, was not only broken but crushed, including the spinal cord. The veterinarian who handled the case felt that there was no hope and did not let her recover from the anaesthetic. He deemed that destruction was the kindest way out.

If it had not been for the crows who gladdened our hearts the next morning I think our household would have mourned the serval cat's passing for a very long time. Cocoa, our old pied-crow patient, easily recognizable by his slightly tilted wing and broken primary feathers, appeared on the grass in front of the verandah while we sat at breakfast. In full view of us he strutted up and down as if to gain our attention, then fluffed up his feathers and arched his back calling until Gail and Guy threw out a plateful of his favourite food, bread and cheese rind. The crow pecked at the crumbs for a moment, then stopped, as if remembering something. Turning away from us, he faced the tall gum tree next to the house and called down a flock of crows who, at his signal, descended on to our lawn and began to feed.

'Look, there are twenty-four,' Gail cried with excitement, rushing off to find some more food. The birds, shyly at first but gradually gaining confidence, advanced to the verandah steps, a few yards from where we sat, to take the crumbs that the children threw out.

'Clever birds,' Toni said. 'They know they are safe.'

He had often told us about the pet rook he had kept as a child and of its intelligence. Crows and ravens were practically the same.

'A grateful patient?' I mused.

The spectacle of the returning crow with his flock of friends seemed stranger than fiction, but then we had witnessed so many remarkable traits and incidences among our animal patients. The depression I had felt at losing the serval cat suddenly lifted. She was beyond pain and distress and we must not mourn the past. Whatever the future held, I could think of no more wonderful way to start the day than facing a lawnful of crows in the early morning sunshine.

Chapter Fifteen

THE LEOPARD WHO WOULDN'T SAY DIE

At the same time as Bibi Kali, the fierce four-pound leopardess, was sharpening her claws on her foster-mother's belly, an older member of her species who lived in the animal orphanage suddenly went into a series of convulsions. Toni and I were just getting ready for our daily walk when Gail answered the telephone.

'Daddy, it's for you, come quick,' I heard her call. 'Samuel says it's urgent.'

'One of the leopards at the orphanage,' Toni explained; 'big trouble, I think, by the sound of it. It can't wait till tomorrow.'

I dropped everything, asked our maid to give the children their supper, and collected our two medical bags and a blanket which was always useful in an emergency. While I checked the drugs and syringes, Toni gave me a rough idea of what Samuel had said, which amounted to the fact that a large leopard who had been perfectly well had suddenly gone mad and was throwing itself about.

We looked at each other, two minds with one thought: surely it could not be Rabies? Not so suddenly, unless the early stages of the illness had been missed. It was getting dark and there were no outside lights in the orphanage. Even with torches, this wasn't going to be an easy case to diagnose.

When we reached the entrance to the leopard enclosures, Samuel and two scouts were already there waiting for us, guiding us to the back fence with their lights. We parked the car and walked to the gate near the sleeping pens where we found, to our horror, that our patient was the gentle, beautiful leopardess known as Sweetiepie, a great favourite with staff and visitors alike, hand-raised before she had been given to the orphanage by an American wildlife enthusiast. As we approached she gave a low menacing growl instead of the usual welcoming purr, falling against the side of the fence in a sudden spasm, her glazed eyes unconscious. When the convulsion had passed she lay completely limp as if dead, and we wondered if she was breathing her last.

'That's what she does all the time,' Samuel said, most upset. 'And then she will get up again and be much better, till the next one.'

Slowly, after fifteen minutes, Sweetiepie rose to her unsteady feet, lurching as she tried to keep her balance. When she reached us she began to purr, not the soft kitten sound, but a sick rasping throat roll, her head tipping drunkenly to one side.

'What a tragedy,' Toni said, 'and it *would* happen to her. We must inject a sedative and get her into the hospital. If she continues to have fits she will die of exhaustion.'

It was now completely dark. We hoped to inject her through the fence, between convulsions, when she came to the edge of the pen as if for moral support. Injecting her with a dart pistol, both painful and traumatic, would be the very last resort for, above all, we wanted to retain her confidence. The shock of a projectile syringe might unleash fury and aggression, the last thing we wanted when she was already nervous, sick and exhausted. All the same, the leopardess was dangerous and unpredictable and we would have to take every precaution in handling her, especially as we had no idea what caused her illness.

'We'll get her tail through the wire and just stick the needle in,' Toni said, preparing the solution, adding a little ruefully that perhaps he would regret his rash decision before the night was out since the last thing a sane person should want to hold in the darkness was surely a leopard's tail!

'It's the only way,' I said heartlessly, swabbing the vial of anaesthetic and sucking up the liquid; 'we must move Sweetiepie where she can't hurt herself, as soon as possible, even if it takes us all night.'

The leopard, being a truly percipient animal, soon

realized that something unusual was afoot but was not quite cunning enough to realize that it was all for her own good. Toni just managed to get enough into the tail as it swept by him to induce an initial but not a very deep sedation; after this she would not give him a second chance, lying enticingly near but not against the fence, moving away and out of reach whenever anyone moved towards her. It was extremely tempting just to go into the pen with a net and catch her, or even to place a loop round her neck.

'Too dangerous,' Toni said, 'she might attack. Taking chances with a convulsive leopard, or any other leopard for that matter, would be just plain stupid.'

We waited for thirty more minutes but Sweetiepie obstinately refused to come near the fence again. In the end Toni and Samuel levered out a plank large enough to admit a hand which seemed equally risky, for although Toni found it easy to get his hand through the wire it was hard to withdraw it quickly. At that moment the leopard collapsed near us as another convulsion seized her. Toni drove the needle into the buttock muscle, extricating himself in the nick of time as she doubled back, too exhausted from the exertion to move quickly. Twenty minutes later she was unable to rise, though she could still lift up her head.

'Deep enough,' Toni said. 'She is too ill to risk giving any more.'

Two game guards and Toni now entered the pen, armed with an old door for a stretcher and the blanket. Raising her head in surprise at the strange commotion, Sweetiepie permitted herself to be rolled into the blanket and on to the stretcher and lifted on to the tail-board of our station-wagon without any show of resistance. Very slowly Toni drove to the animal hospital with our stretcher-case sitting bolt upright, her enormous eyes reflecting green-gold in the light of our torches; 98 pounds of mature leopard being transported on the back of our car as easily as a domestic cat.

'How are we doing?' Toni called from the driving seat;

'only a few more yards to go.'

'Patient sitting up, but controlled,' I answered; 'might need a little more sedation before we bed her down for the night.'

Once in hospital, electric light made a full examination very much easier. As far as we could see everything was normal, eyes, ears, mouth, body and limbs. No sign of snake-bite or injury, no swellings or abnormalities anywhere. I thought of our miniature leopard at home, not even as large as Sweetiepie's head! Would she ever grow to this size?

The darkly rosetted, soft-furred skin under my hand began to twitch, the muscles to move. Toni reinforced the anaesthetic with a tranquillizer which would hold her asleep and stop the convulsions for at least twelve hours. We injected a wide-spectrum antibiotic and vitamin-B complex and checked her temperature, finding it raised to 104°F., 101 being normal. Even that did not tell us very much, since the activity of a convulsing body always elicits a thermal response which would gradually decline as activity ceased.

There were many possible causes of her illness, the most likely being the inflammation of the brain (encephalitis) and meninges (brain covering) caused by an infective organism, virus or bacterium. She had been vaccinated against feline enteritis long ago but there were new and resistant strains abroad which had already taken toll of other wild felines.

It could be a case of acute poisoning, such as metal, strychnine or prussic acid or damage to the brain from an injury, or tumour causing pressure. Lastly, there was always the dread of Rabies though the symptoms we had witnessed did not tally with hydrophobia, as Rabies was often called, when the behaviour pattern changes either to aggressive or an opposite, fear, syndrome, always accompanied by the inability to swallow.

Rabies meant mortal danger to any human who had been in contact even by means of the smallest, often unnoticed wound through which the infected saliva could penetrate. If there was any doubt then the animal in question was destroyed, the human contacts were protected with serum while pathological tests on the suspected brain were in process. After that, if the result was positive, a series of Rabies vaccinations were injected, a painful and dangerous process which I had undergone years ago when I had unknowingly grappled with a rabid dog.

Toni and I were in full agreement that this was not a case of Rabies and took full responsibility for the leopard's care. We wrote a note to be delivered to the Warden first thing in the morning giving our report, promising to return in the lunch hour of the next day, unless anything unforeseen occurred before that. Our patient, now fully relaxed, was placed into a box cage with instructions that she must be watched over most carefully and given water to drink as soon as she awoke. It was pathetic to see her confined in such a small area, yet at this stage it was the safest place. Above all, her tail was easy to grasp through the wide mesh so that treatment would not be as hazardous as it had been when she was in her pen.

Sweetiepie's illness lasted for many months and her life swung in a precarious balance most of the time. Gradually the frequency of her convulsions decreased, but she retained the drunken 'sailor-walk' and the one-sided head-position. She took water, ate ravenously as animals often do when affected with encephalitis, and behaved rather unpredictably probably because she had, or so we imagined, an almost constant headache. She remained in her small cage in the orphanage hospital and we continued to treat her with a number of different kinds of medicines, keeping her slightly tranquillized all the time. A week after her first symptoms had appeared we took her to the veterinary clinic to be X-rayed in case something like a tumour of the brain might show up. The X-ray plate showed nothing abnormal though even this did not exclude a growth of some kind, since, unless very dense, it would be difficult to demonstrate.

It spite of constant therapy Sweetiepie's imbalance and sudden periods of collapse did not improve, though her acute convulsions gradually disappeared. As the weeks went by many of her admirers lost hope, wondering if continued effort was justified humanely or economically. She was a large, unusually fine-coated animal and her skin would fetch a large sum of money. Although mature she had not, to date, bred with her male counterpart and it was thought that perhaps she was sterile.

Whichever way one looked at it, her continued existence was neither essential nor useful and her space in the hospital could well be used by others. The Warden, a kindly man who had permitted us to continue treatment though there was so little hope, reasonably enough began to mutter about the value of the leopard's skin, which would certainly deteriorate with prolonged illness. The orphanage was chronically short of money and the skin would fetch a good price, yet here we were, throwing good money after bad with daily injections and special feeding.

After two months of alternating medication and vitamin therapy and several consultations with other clinicians, we decided to try and make one final attempt at diagnosis. A lumbar puncture might be the way to assess the state of the cerebrospinal fluid which our friend and physician, Doctor Blaine, who had great interest in the case, would take to be analysed at the Pathological Laboratory. Since her illness seemed to have reached a chronic level, Toni and I had to accept the fact that if this last test led to nothing, our beautiful leopardess would have to be destroyed.

Once again Sweetiepie was anaesthetized, placed on a stretcher and put on a large table. Toni, thinking of the pelt-collectors, clipped the hair along the spine from top to bottom so that the skin, at least, could not be an incentive for destruction. After sterilizing the area Toni inserted the needle between two lumbar vertebrae, not certain of the procedure since neither he, nor anyone else for that matter, had ever carried out a leopard lumbar puncture before. In the end we only managed to withdraw a little bloodstained fluid, useless for examination since the protein balance would be upset by the blood cells. For the sake of our deeply slumbering patient we decided to call it a day fearing to prolong the anaesthetic time yet once again.

'Give her a few days to recover from that,' we asked the Parks warden and Samuel Ngethe, 'and then we can finally reassess the situation.'

After three days, the time estimated for complete detoxication of the anaesthetic, Sweetiepie's uneven gait began to lose its lurch. She was still confined to her small pen and it was difficult to be certain, yet within one week her one-sided head carriage, which had been a constant feature of her illness, began to normalize.

At first we would not allow ourselves to believe that this new state of affairs was anything but temporary, expecting the dreaded symptoms to return at any moment. We saw our patient twice a week, maintained vitamin therapy and special feeding and alternated between thanking our Maker for her improvement and blaming Him for giving us false hopes when we knew, scientifically speaking, that it was quite impossible for her to make a recovery.

Three weeks after Sweetiepie first showed definite signs of improvement we went on overseas vacation, leaving her in the able hands of our colleagues. By this time she was well enough to return to her own enclosure and was later on moved to larger quarters in the new animal orphanage, confounding everyone not only by her startling return to radiant health but by becoming pregnant at long last.

'Remarkable, inexplicable, unpredictable, just like a female,' was the consensus of opinion, which was shortly to be reinforced by yet another, most unexpected act in the drama of the enigmatic leopardess.

Although she seemed perfectly contented in her new home and as even-tempered as she had been before her long illness, Sweetiepie, who had beaten death from her own door, now committed an act of murder. Before her own cub was twenty-four hours old, she destroyed it with one savage snap of her huge jaws as if it was no more than a large rat that had strayed into her domain. After this act she was perfectly serene, ignoring the immobile body as if it had never existed, once more alone and lonely in her captivity, when motherhood and companionship had been so nearly within her reach.

Had she been at a loss as to what to do with the blind, wriggling furry intruder, her maternal instinct not yet fully in play? Or had her deepest sub-conscious, perhaps, been aroused by the act of birth, that sub-conscious part of her which longed to be free and revolted at her useless, sterile existence expressing itself in a sudden act of violence towards her own young, also condemned to be a life-long prisoner?

Had she been wild, she would have borne her cub alone, in some secret lair, and raised it without interference from man or beast. Here, in the noisy, public zoo there was no hope of privacy or silence, no respite from the prying eyes. Had the leopardess been free her cub might have lived, and yet, at the rate man hunts down her kind for the glossy, spotted skin, only the caged ones, like Sweetiepie, might eventually survive the slaughter.

Chapter Sixteen

THE WIRE NOOSE

Within East African Parks and reserves the wardens were fighting a lone and relentless battle against the destroyers. With inadequate staff and far too few vehicles to patrol their vast terrain which gave access to unprotected land either as hunting blocks, habitation, agricultural land or public highways, their greatest problem was that of poaching. No longer done only on foot, it had developed into highly organized robbery, effected with speed under cover of night or in the densest, most inaccessible places. The rhinoceros was the most sought-after prize, then the elephant, the giraffe, buffalo and larger antelope. Sometimes the poachers' plans were foiled and they were apprehended by the Law, but even in such a case their victim had usually already been injured or killed.

Worse than killing with gun or spear was the setting of traps: thin wire or reed snares for small antelope, thick wire nooses for the larger game placed along well-used game paths, so that the victims stumbled into them. Because the poachers set a great number of these, they were unable to patrol them regularly. Thus the animals caught died a slow death of unutterable, indescribable agony, eaten alive at times before they were dead, as in the case of a beautiful spiral-horned kudu Toni and I saw on an anti-poaching safari in sanctuary land. He must have walked into a trap some days before, judging by his condition, his front right leg twisted beyond belief as he had tried to escape from the noose that still held his hoof. Nothing in the world could have saved the leg of that agonized antelope, though he might have lived on for days had the *coup de grâce* not been delivered at once.

Commercialized, modernized poaching was Ted Goss's most pressing concern. He had recently taken over the beautiful Tsavo Park West from the retiring warden, Tuffy

Marshall, and had found his half of the Tsavo Park complex, which consisted of 5,000 square miles on the west, as varied with vegetation as it was beset with worries. A mixture of dense bush, forest, riverine landscape and mountainous terrain, the Park lies on the south side of the Nairobi–Mombasa road, containing the world-famous Mzima Springs, where fifty million gallons of clear highland water bubbles down daily from the volcanic soil of the Chyulu Hills, supplying water to man for hundreds of miles.

Ted, young and energetic, was well used to all the worries and complexities which a Park's warden had to meet. He had put the Meru National Park on the map, had fought and won the three pairs of white rhinoceros for the Park which had been imported from Zululand for the purpose. The rhino, his personal responsibility, had offered a tremendous challenge, for although they had long ago existed in that part of East Africa, the imported six had to acclimatize to new land, new diseases and new stress.

Ted had always been a pioneer at heart: when the rhino had settled down, he decided that it was time to learn more about his other animal residents and began with the elephant. During an immobilization operation he was critically injured but miraculously survived. After one year on his back and many trials and tribulations to try his strength, he made a complete come-back, taking over the much-coveted and not yet completely developed West Tsavo National Park. There was so much to do that it was difficult to embark on scientific wildlife research, though elephant 'exclosures', land unexposed to elephant, were already planned so that at last the enigma of the elephant's eating habits and the toll he took of the vegetation could be solved.

Whether research, or rescue of injured animals, Ted left no avenue unexplored. In his Park help was given when it was needed at whatever sacrifice, not for five days from eight to five, but for seven days a week and for twenty-four hours of the day.

The snaring and poaching of the large game animals,

such as the black (hook-lipped) rhinoceros, infuriated Ted Goss more than any other crime committed against his Park. Already he had darted several which had been speared in preparation for the slaughter and the removal of the precious long anterior horn. Others were shot with large-calibre modern firearms, the poachers using fast light vehicles to make their get-away. Ted fought and begged for sufficient funds to assemble more anti-poaching vehicles equipped with radio-transmission, but money was hard to come by and there was never enough to meet all his needs. Not only his but those of every warden in every Park and reserve in Kenya. The tourists came and went, sending glowing reports of the wildlife they had seen across the world; they would have been astounded and horrified to know that each day a fierce battle for the continued existence of the animals they so enjoyed was being fought under their very noses! With totally inadequate funds and little money in the kitty for rescue work (£50 a year in one case!) the wardens and their staff carried on as best they could. Occasionally, when an injured or trapped animal posed a particularly difficult problem, Toni and I were asked to assist – as in the case of the reticulated giraffe.

A scout on patrol had sighted the injured animal and had reported to Ted. It was a female giraffe, part of a large herd in the south of the Park, her neck caught in a wire noose. Ted, well trained in the art of immobilizing, hesitated about darting her. Giraffe were so much more difficult to deal with than rhino, or antelope or zebra, by virtue of their shape and their physiological make-up which did not tolerate large drug dosage or the shock of being 'knocked down' as the immobilizing drug took effect.

A snared giraffe, already stressed by a restricted air intake due to the pressure of the wire on the neck, would have to be caught with the utmost care.

Toni and Ted had often worked together in the past and knew each other's worth. Ted recognized Toni as a highly skilled veterinarian and an expert in his field, so that when Ted made a dig at scientists, which he frequently did, he knew and we knew that Toni was exempt. My own status, in Ted's eyes, seemed to vary according to circumstances: he had little respect for women in society and certainly did not rate them as equals, or even near-equals. Yet when I took up my medical case and got to work, Ted accepted me not only as an equal but as a professional. I often detected the bafflement in his eyes when I once more became a normal domestic female, an apron tied round my waist, waiting on him in my home. I think for him I wavered between being an inferior female and a sexless vetess. There seemed little hope that men like Ted would ever recognize the fact that a professional female can be efficient and yet retain her sex appeal!

Our home was in mid-morning turmoil when Ted arrived. Toni's wildlife physiological research charts were spread all over the table, and these were being discussed with his PhD student between cups of coffee and telephone calls. Sam McGinnis, the biotelemetric instrument maker and biologist who had arrived from California to work at Toni's research centre, was vainly trying to take a photograph of his small recalcitrant son with our mongoose and genet which hung on to my shoulders for dear life. Each time the toddler came too near, his back crest went up and his ears went down, accompanied by genet growls which increased in intensity as the sharp claws dug into my flesh. Toni, flinching at the mixture of screams, howls and growls, tried his valiant best to be polite to everyone though I could see from the hunted expression on his face that he might bolt at any moment.

Ted, ruddy-complexioned, cool and composed in his smart fawn bush-jacket, looked on with a benign, amused expression on his face, obviously enjoying the chaos in someone else's home!

'Can you come down day after tomorrow to catch this giraffe?' he said, finally, when the mongoose–genet commotion had died down and Toni's student had departed. 'I'm not even sure if we can find her again, she might well be dead by now,' he added, 'but I'll be grateful if you could both come and take the chance.'

That had been Wednesday. The earliest we could leave Nairobi was Saturday, which meant working on Sunday, and making the three-hour return journey at dawn on Monday. Providing nothing desperate occurred in the next two days we would be there latest Saturday by dusk; on second thoughts, apart from a major disaster, what could possibly be more desperate than a snared giraffe?

Ted and Else's home, only a few miles from the Mtito Ndei gate on the Nairobi–Mombasa road, had been built on top of a rise facing the undulating mist-grey Chyulu Hills. We arrived just in time to see a group of five elephants emerge from the dense acacia and baobab woodland to take their sundowners from the water reservoir below the house. The azure sky, speckled with clouds back-lit by the golden-red electric evening sun, slowly deepened into dark blue as slanting shafts of last light momentarily touched the elephant-grey hills until the radiant evening star announced the coming of night.

I noticed that our bedroom window was newly barred and remembered Else's warning last time we had stayed there.

'Do remember to close them,' she had suggested gently. 'The lion still comes to drink from our bird-bath in the front garden.'

Else and Ted had small children but they somehow seemed to know which way they had *not* to go. Wild cats large and small, elephant, buffalo and practically all the inmates of the Park came to visit them and leave their tracks and traces at some time or other.

It meant taking special care and a life without many human playmates, but the children thrived and grew into a rare awareness of Nature which city children sadly have no chance to comprehend or enjoy.

Sunday morning early was set for the giraffe rescue operation, that is provided the giraffe had been found. It took an hour to reach the area deep in the south, the most inaccessible section of the Park, where the team of scouts and the assistant warden, Bob Guyio, were waiting for us. They had sighted the snared animal among a herd of 15 and were slowly following them as they moved from tree to tree in a valley below the Taita Hills.

Beyond was a thick woodland into which it would be hard to follow, but to the south lay more open country, though there the ground was strewn with lava boulders, often hidden from view by long grass.

Of the two, we preferred the rocky, more open ground and slowly headed the herd towards it away from the trees, until we were able to get a clear view of the female. The wire snare tightly encircled the base of her neck, while the free end hung down between her forelegs, trailing to the ground as she ran.

'Let's head her off from the herd and then take a shot,' Toni said. We were at least four hundred yards from the nearest giraffe, a dark brown bull who seemed to stay near the injured female as if protecting her. He was beautifully marked, his reticulations clearly defined as if the pattern had been chiselled into his hide. His long powerful neck stretched proudly upwards as he cantered a little behind the rest of the herd.

The female was a great deal lighter in colour and looked feminine, almost fragile, yet even in her stressed condition she possessed the startling beauty of her kind, her body hardly touching the ground as she moved in and out of the trees, her black switch flowing behind her.

How long, with the cruel wire cutting more and more deeply into her flesh, would she live if we did not manage to immobilize her that day or the next? It was only a matter of time before the swinging cable became entangled in some thick thorn shrub, condemning her to a lingering death.

At last we passed her as she began to lag behind the others, her stamina much less than theirs. They loped on over the hill, leaving her far behind as she became more and more distressed, though she still attempted to shake us off among the rough boulders.

I remembered a dead giraffe I had seen not so long ago in another Park which must have fallen prey to lion or leopard. Nothing had remained of it but bones, hair and skin which had been scattered far and wide among thorn scrub and tufts of grass. It must have died very quickly, for the cats mete out a merciful death, not the slow cruel murder that man thoughtlessly commits, such as the slow strangulation of a snared animal.

Suddenly our quarry stopped, turning back to look at us with intense concentration as if trying to assess our intention. Then, after a few moments of rest, she cantered on, making good speed in spite of her predicament.

'We'll have to chase her, after all,' Ted said. 'She's very lively. You ready?'

'Ready,' Toni said.

He had prepared three darts (two spares just in case) as soon as we had first spotted the giraffe. Each had been filled with his famous 'cocktail' mixture: the knock-down drug, the sedative to guard against any excitement the first substance might elicit, and a tranquillizer which would prevent a fear reaction after the first substance was reversed at the end of the operation. All this was contained in only two cubic centimetres, which meant that the dart could be light and short, not long and heavy and therefore as traumatic as immobilizing syringes had been in the days when Toni had first pioneered the field.

I watched him as he sat, gun in hand, his face intent and determined, his whole being gathered up for the effort he was about to make the moment Ted caught up with the giraffe. Toni had practised very little of late with either dart gun or crossbow. Having perfected the method and the drug, having opened up fields of scientific conservation and wild-animal rescue work which no one had imagined in their wildest dreams as recently as twelve years ago, he had then stepped back and embarked on a new kind of wildlife research, deeming that others were now as able as he to put immobilizing methods and drugs into practice.

'You cannot escape,' I often warned him as he grumbled over the endless requests for advice in wild-animal capture which still flooded his mail basket. 'Once you have achieved a new field, and especially now that you have published a book, people will continue to ask you for help for as long as immobilization is a feature of conservation.'

'The Flying Syringe' – the story of how and where and why all this had come about had just left the press. At last much of the misinformation regarding wild-animal immobilization might be dispelled and the unskilled amateurs and would-be immobilizers might cease to waste the lives of wild animals, such as the rhinoceros, many of which had died needlessly in eastern Kenya not many years before.

'Okay, we're just about right if you want to take a shot,' Ted said.

Toni, in the front seat next to him, rose to his full height through the padded open roof and took aim as soon as he had the giraffe's rump in his sights. To *me*, looking on from the back seat, I was only conscious of a blur of hooves as the galloping form of fawn on white plunged and jinxed twenty yards in front of us. How Toni held his target steady, even for a moment, I shall never understand, for the Land-Rover, bumping and sliding wildly over the rough stony terrain, made a very unsteady aiming platform.

'Voom.'

It was a small sound compared to the straining Land-Rover engine and I would have missed it but for the second-long silver flick as the dart found its mark on one side of the tail, embedded firmly in the muscle. We slowed down to allow the harassed giraffe time to get her breath. Through our binoculars we could still see the bright orange tail of the flying syringe clinging to her buttocks where it was likely to stay, for most darts which are rejected fall out on impact in the very first moments after projection.

'Good shot,' said Ted, relieved and pleased.

'Thank the dear Lord,' Toni answered, 'I am so out of practice that I can claim no credit whatsoever for hitting the bullseye. But we've only begun,' he added. 'Now to catch her and put her down.'

Giraffe are unlike any other kind of wild animal from the immobilization point of view. Toni had already discovered that no two animals react in quite the same way, whatever species they belong to. Even animals of the same kind vary according to season, breeding periods, and individual temperament. The nervous creatures, such as the black rhino, are very much more difficult to capture than the more phlegmatic animals like the eland and kongoni.

The secret of success with giraffe was to drug them just sufficiently to slow them down so that they would walk into an extended rope which brought them to a full stop. After this, without panic or struggle, they would allow themselves to be gently collapsed.

Toni's weight estimation and dosage level were perfect. After twenty minutes the giraffe reduced her pace, panting with open mouth each time she rested. The dark-patterned male, her consort, had again left the herd, making valiant efforts to entice her back in spite of our noisy manœuvres. But the female seemed hardly aware of his presence now, moving in short bursts with the typical gait of a morphinized animal, head tilted upwards and the pastern joints flexed at each step like a circus pony.

'The rope!'

Ted had explained the catching technique to his team, but the rope was slippery and the slow amble of the giraffe was still equal to the fastest human pace. At last, after a few unsuccessful attempts, she stood still as the scouts doubled the rope from back to front, her legs bunched together like matchsticks. In the distance the herd browsed peacefully on the high acacia, the male apparently no longer interested in the fate of his unresponsive mate.

Toni and Ted leapt out of the Land-Rover and directed operations while I assembled everything we needed on a metal tray. The medical case was ready: syringes, wound dressings, antibiotics, shock medicaments. Toni already had the antidote syringe and needle in his pocket; Ted went for the wirecutters the moment the giraffe went down. It had taken very little effort to tip her off balance as the team of game scouts gave a strong tug on the ropes to tighten them further and put gentle pressure against one side of her body. Without injuring herself, or any particular sign of distress, she sank on to the ground, her heavy head supported by four scouts.

This was one of the problems whenever a ruminant animal was immobilized: if the head was lower than the body there was always danger of regurgitation when the ruminal contents might flow into the lungs. If this happened there was little that could be done; the consequence was usually death from inhalation pneumonia. In all cases speed of action and division of labour were vital so that the animal could be up and away within half an hour, or forty minutes at most.

While Ted unsuccessfully tried to cut the wire noose, I injected penicillin into the neck where the skin was so tough that I had to find the widest-gauge needle I had and even this was bent by the resistance of the hide.

'It is a protective feature,' Bristol Foster, the ecologist who had studied giraffe in the Nairobi National Park, had told me. 'Since love-play and fighting are enacted with the neck, a specimen with anything but the thickest, hardest skin will not survive and therefore the weakness will not be passed on.'

It was not only tough but seemed to be reinforced with a kind of innerspring elasticity which probably blunted the needle as it drove through and into the muscle. I injected a large dose, one which would take three days to be absorbed. If infection had set in, this should arrest it.

'Can't cut the damn thing!'

Ted was frantically trying to snap the wire, four strands thick, but was meeting with no success at all. He ran back to the Land-Rover to try and find a hacksaw but meanwhile the scouts had managed to unravel the high-tensile fencing-wire snare, four strands intertwined and therefore extremely strong. By the time he returned the neck had been released and I was applying a mixture of fly-repellent and antibiotic paste, bright yellow and very potent, working it into the damaged tissues as deeply as I could.

'Hold the head higher,' Toni commanded as he sucked the antidote into the syringe. The scouts were sweating with the effort of supporting the front end of the giraffe which hardly moved throughout the whole procedure. We watched her reflexes, her breathing, but there was no cause for alarm. She was taking it all very much in her stride.

'Off with the ropes.'

Toni was ready to inject into the jugular vein while I handed the ointment and syringes to Hanne, Ted Goss's sister-in-law who made a marvellous assistant. I put pressure on the vein high up in the jugular groove at the top of the neck and almost at once the blood vessel bulged, an easy target for Toni's needle. The ropes were off, everyone was out of the way. Toni injected, once into the vein and a few seconds later into the shoulder muscle. Within one minute the giraffe responded, swinging up her head and neck in an effort to get to her feet: first very groggily, the second time more strongly. We stood by in case she needed help, for it seemed such an immensely long way up to her fifteen-foot head. But she had no need of us. As the antidote counteracted the drug effect she suddenly, at the third effort, swung up on to her feet, looking down at us with a surprised, but not unfriendly expression in her doe-like eyes. Again, as on many previous occasions, I noticed how superbly long were her eyelashes and how exquisite her sensitive soft-muzzled face. The tranquillizer Toni had added to his initial 'cocktail' had successfully removed any fear she would normally have felt if suddenly cornered, without the slightest knowledge of how she had got there. It would have been a frenzied fear which could turn to panic, aggression or self-injury as the animal took flight.

I had so often before witnessed the lightning recovery of an immobile animal and yet each time the transformation from limp inactivity to vibrant life amazed and startled me anew. I watched the yellow-collared giraffe lope away from us with even, unhurried rhythmic grace, freed at last of her lethal burden, her gait no longer hampered by the inhibiting drug. Anxious to take one last photograph for the record, Toni and Ted drove off in pursuit of the receding form, clearly outlined against the upward slope at the edges of the woodland.

Suddenly a large black cloud drifted between earth and sun, already dipping towards the vivid-green Tsavo horizon. I found a tree stump and from its added height tried to catch one last glimpse of the speckled, diminishing shape, but it had melted into the shadows that the cloud had drawn, like a veil, over the trees and bush and the eroded rocky grey-brown volcanic earth.

One more job done successfully, one more wild animal safely rescued; yet out there, in the vastness of Africa, countless wild creatures were lying injured, starving and dying, victims who would never be reached or rescued. I turned back to the game scouts who were helping to gather our scattered belongings. They were tired but happy, even jubilant that our day had gone well, for only rarely did a poacher's victim escape with so little damage. Had they always cared for the fate of the wild or had they gradually learnt to care as had so many others who lived and worked in the glorious last strongholds of Nature? Many of the most ardent hunters, not only in Africa but throughout the world, had turned into devoted conservationists; some of these the very men who were now the guardians, the wardens of the great wildlife sanctuaries. Respect, compassion and admiration akin to love for natural beauty in all its shapes and forms had gradually crept into their hearts until the snuffing out of life, other than as an act of mercy or to protect human life, had faded into a memory of the distant past.

To listen to the wild, to be among them, not for a day but for many days and nights, camped by a cool bubbling mountain stream, or at the foot of a rainbow-edged waterfall, is the greatest tonic man can take against the relentless deadly onslaught of the noise-ridden, polluted world. Let him not go as a stranger or as an aggressor, but as a humble observer. However intense his own turmoil, he will be caught up in the spell that the timeless wilderness casts about him, unburdening him at last of his troubled, confused ego, so that he will become as a grain of sand in the desert, or a feather on the eagle's wing, an anonymous, but essential link in Creation's plan.

EPILOGUE: WHAT HAPPENED TO OUR WILD PATIENTS

Once our patients leave our care, scattered to the four winds as they are, it is often difficult to follow the course of their lives, especially if they are released into the wild.

There is so much we would like to know about them: their adaptation to new surroundings, new companions, new challenges. And if they have a permanent handicap, like the genet with the injured jaw, how are they able to manage in a semi-free existence? To know the outcome of our treatment gives us valuable scientific information, for often it is only after months that healed, recovered tissues are thoroughly tested.

We have been fortunate with the group of wild-animal patients described in this book. News items of their welfare and whereabouts have been reaching us over the past months, at the time this book goes to press.

Alicat the genet (renamed Spitpuss) was adopted by a family on the edge of the Aberdare National Park and could not have found a more welcoming or loving home. When I saw Mrs Thatcher she was a little anxious because Ali had been absent for some days, rather in the way he gave us the slip while his jaw was still fragile. She told us that in spite of his handicap he was actually catching small prey animals and was coming and going as he wished, though remaining as tame and mischievous as he had been when under intensive care in our home and previously, in Manyara Park. His pal, the mongoose Wiji, had been given to another animal lover farther north and was often taken on safari along with her clients and other pets. Wiji, though only just beginning to recover when she left us, was now completely fit and back to her old rotund shape.

Poppet, the genet we raised, lives on, semi-free, in her large enclosure on the farm in the northern highlands of

Kenya, a male genet her companion by day, a tree hyrax by night. We had all hoped for some progeny but so far there has been no sign of romance. Meanwhile, we are searching for a suitable location where both genets can be released: a genet area but not too thickly populated with genets. We cannot forget the onslaught Alicat suffered from the territorial male in Manyara Park.

Lolita, whom we saw on a recent visit to the Sheldricks in Tsavo Park East, had grown immensely beautiful and, if possible, even more feminine than before. Her eyes were as large and limpid as ever, her sensitive ears eternally mobile. Our infant duiker of yesterday had metamorphosed from a skittish adolescent to a self-possessed, fully-grown adult. Somewhere, beyond the edge of the garden, shy and out of sight, a wild male duiker hopefully hovered in attendance, waiting for Lolita to respond to his call.

Wiffles, though hardly a patient of ours, cannot be left out of this epilogue. Not long after Lolita arrived in his domain he decided to move on, though only a few hundred yards down the Parks' road, using the Glovers' back garden as a kind of refuge, a place to come back to, but hardly more. Wiffles had found a mate, probably for life, as is the habit of the dik-dik, but as yet she kept away from human habitation. Perhaps later she would allow herself to be wooed by Wiffles's human friends.

Gilka, when we last heard, was progressing reasonably well. The nose was still bulbous, but the swelling was no longer increasing in size. With medication it might disappear altogether.

Boy, alas, revelled in his fastness of the north-west for only nine months, having gradually ventured farther and farther afield from the main release camp. Tragedy struck when he attacked one of George Adamson's servants, the only person to whom he had ever shown animosity, and who perhaps had failed to follow George's dictum: that rehabilitated lions must be treated as truly wild. With one deft shot George snuffed out the life of his beloved lion, circumstance forcing him to take the split-second decision which he could never reverse. Though baffled and heartbroken at the time at this sudden cutting-off of such a meaningful life, George continues his marvellous work with a young group of lions.

Zorba, our adored, delinquent cheetah, died almost a year after he returned to the Animal Orphanage. Toni and I had been away on leave and when we returned we could not find him or his two companions. Finally we learned that all three had perished in one night without any history of previous illness. They had almost certainly been poisoned; exact cause unknown.

Manyaka, our ancient squirrel of joy, died of old age one cold winter's day, having confounded our predictions five autumns running that his time to leave us had surely come. Often, as I sit below the swinging branches of his tree-bush, I seem to hear his staccato chirrup, welcoming us to his bower where he entertained us so well with his aerobatics.

The crow, Cocoa, who brought his flock to feed on our lawn, was last identified for certain one year ago. He hopped into our parsley patch at the kitchen door, caw-cawed loudly and demandingly and immediately pitched into Jessie's food bowl as if he had never been away. Apart from his familiarity with our household, he could easily be identified by the slightly tipping wing, the only remaining evidence of his injury.

Hugo the chimp has grown enormously since he left our care. When we visited him, very recently at his home outside Nairobi, he took a little time to realize that we were old friends. At first he stared at us in frowning, puzzled silence, then suddenly recognition lit up his sorrowful, enigmatic face. With a broad toothy smile and amidst many delighted uuuguuugs, he went first to Toni – the dominant male he remembered so well – clinging to him as to a long-lost friend.

Then he climbed on to my stomach in the accustomed way, and together, for some ever-remembered moments, we celebrated our reunion, his calls more mature now and very definitely male. His chest muscles had developed well and the fragile razor back was completely covered with firm hair-covered flesh. Yet his face still held that haunted look, that sorrowful tragic air, as if he was always conscious of the world he had lost for ever.

The little leopardess, Bibi Kali, died some months after she was born, unable, we suspect (since we were absent), to overcome the lung infection which she had contracted but recovered from even before her eyes were opened, and which must have lain dormant in her body as she grew.

In contrast, Sweetiepie, the leopardess who would not die, lives to this day, though she is still without cubs.

And what of the giraffe, whose cable-snare has joined the Warden's collection? Almost certainly she survives, for her group is often seen, though never at very close quarters.

And Jessie and Mr Seb? They continue to thrive – glad of our undivided attention when we have no animal patients, yet just as pleased when we have visitors who have need of care. I cannot imagine our kind of life without them.

INDEX

Index

Animal Books in Fontana

JOY ADAMSON

Born Free (illus.)
Forever Free (illus.)
Living Free (illus.)

PHILIP BROWN

Uncle Whiskers

GERALD DURRELL

Beasts in My Belfry
Birds, Beasts and Relatives
Catch Me a Colobus (illus.)
Fillets of Plaice
Rosy is My Relative
Two in the Bush

JACQUIE DURRELL

Beasts in My Bed (illus.)
Intimate Relations (illus.)

HUGO VAN LAWICK

Solo (illus.)

PHILIP WAYRE

The River People